Learn Python Programming

A Practical Introduction Guide for Python Programming. Learn Coding Faster with Hands-On Project. Crash Course

Jason Crash

© Copyright 2021 – Jason Crash - All rights reserved.

The content contained within this book may not be reproduced, duplicated or transmitted without direct written permission from the author or the publisher.

Under no circumstances will any blame or legal responsibility be held against the publisher, or author, for any damages, reparation, or monetary loss due to the information contained within this book. Either directly or indirectly.

Legal Notice:

This book is copyright protected. This book is only for personal use. You cannot amend, distribute, sell, use, quote or paraphrase any part, or the content within this book, without the consent of the author or publisher.

Disclaimer Notice:

Please note the information contained within this document is for educational and entertainment purposes only. All effort has been executed to present accurate, up to date, and reliable, complete information. No warranties of any kind are declared or implied. Readers acknowledge that the author is not engaging in the rendering of legal, financial, medical or professional advice. The content within this book has been derived from various sources. Please consult a licensed professional before attempting any techniques outlined in this book.

By reading this document, the reader agrees that under no circumstances is the author responsible for any losses, direct or indirect, which are incurred as a result of the use of information contained within this document, including, but not limited to, — errors, omissions, or inaccuracies.

Table of Contents

Introduction 9

Chapter 1: Python 14

    Why Should You Learn Computer Programming?....14

    Advantages of Modern Programming 17

    Why Did Python Emerge as a Winner Among a Lot of Other Computer Languages? 21

    What Is Python (and a Little Bit of History) 22

Chapter 2: Importance of Python 28

    What Can You Do as a Python Programmer? 28

    Example Program to Show You a Good Overview of the Python Programming 36

    Python Reserved Words 40

Chapter 3: How to Install Python 41

    Installation and Running of Python 41

        Official Version Installation 41

        Other Python Versions 45

        Virtualenv 46

Chapter 4: The World of Variables 47

Chapter 5: Data Types in Python 55

    Basic Data Types 55

Tuples and Lists 58

Dictionaries 60

Chapter 6: Operators in Python 62

Mathematical Operators 62

String 63

Comparison Operator 64

Logical Operators 65

Operator Precedence 68

Chapter 7: Execution and Repetitive Tasks 69

If Structure 69

Stand Back 73

If Nesting and Elif 76

For Loop 78

For Element in Sequence 79

While Loop 81

Skip or Abort 82

Small Exercise to Review What We Learned Until Now
.... 84

Chapter 8: Functions and Modules 86

What Are Functions? 86

Defining Functions 89

How to Call Functions? 92

Function Documentation 94

Parameter Passing 97
    Basic Pass Parameters 97
    Pass the Parcel 100

Unwrap 103
Recursion 105
    GAUSS and Mathematical Induction 105
    Proof of the Proposition 108

Function Stack 108
Scope of Variables 110
inner_var() 111
Introducing Modules 113
Search Path 115
Installation of Third-Party Modules 118

Chapter 9: Reading and Writing Files in Python 119

    Storage 119
        Documents 119
        Context Manager 122
        Pickle Pack 126

Chapter 10: Object-Oriented Programming Part 1 130

    Classes 134

Objects 136

Successors 141

    Subclasses 141

    Attribute Overlay 143

What You Missed Out on All Those Years 145

    List Objects 145

    Tuples and String Objects 147

Chapter 11: Object-Oriented Programming Part 2 148

Operators 148

    Element References 152

    Just a Small Example for Dictionary Datatype ... 153

Implementation of Built-In Functions 153

Attribute Management 154

    Features 158

    getatr() method 161

Dynamic Type 163

    Mutable and Immutable Objects 165

    Look at the Function Parameter Passing from the Dynamic Type 168

Memory Management in Python 169

1. Reference Management *169*
2. Garbage Collection *170*

Chapter 12: Exception Handling 173

What Is a Bug? 173
Debugging 176
Exception Handling in Detail 178

Chapter 13: Python Web Programming 181

HTTP Communication Protocol 181
http.client Package 187

Conclusion 188

Introduction

Congratulations on purchasing *Learn Python Programming*, and thank you for doing so!

The following chapters will discuss Python programming in detail, with a well-versed example that will help you get a better understanding of different programming concepts with the help of Python. You've taken the first step to learning a programming language that is famous for its robustness and simplicity.

Taking Python as an example, this book not only introduces the basic concepts of programming but also focuses on the programming language paradigm (process-oriented, object-oriented, function-oriented), as well as the programming language paradigm in Python. This way, the reader not only learns Python but also will have an easier time learning about other programming languages in the future.

The appearance of computer hardware performance has developed by leaps and bounds. At the same time, programming languages have also undergone several changes, resulting in a variety of programming paradigms. Python, with its simplicity and flexibility, has made its way to software industries in spite of many programming languages. Throughout history, we have experienced not only the features of Python but also the main points that the language is meant to address.

Computing has a long history dating back to thousands of years ago. People can calculate and remember—but what's even more remarkable is their ability to borrow tools. Humans have long used methods and tools to aid in highly complex cognitive tasks such as computation and memory. By tying knots in ropes to record cattle and sheep in captivity, our ancestors had long been able to use the abacus at dizzying speeds. With the development of modern industrialization, the social demand for computation is more and more intense. Taxes need to be calculated, machines need to be built, and canals need to be dug. New computing tools are emerging. Using the principle of a logarithm, people made a slide rule. The slide rule can be moved in parallel to calculate multiplication and division. Charles

Babbage, a 19th Century Englishman, designed a machine that used a combination of gears to make highly accurate calculations, hinting at the arrival of machine computing. At the beginning of the 20th century, there were electromechanical computing machines. The electric motor drives the shift gears to "squeak" until the calculation is made.

During World War II, the war stimulated the need for computing in society. Weapon design requires calculations, such as the design of a tank or the outer hull of a submarine trajectory. The militarization of society requires calculations, such as train scheduling, resource allocation, and population mobilization. As for the missiles and high-tech projects like nuclear bombs, they need massive amounts of computing. Computing itself could even become a weapon. It's worth noting that it was Alan Turing who came up with the idea of a universal computer theoretical concept for the future development of the computer to make a theoretical preparation. The top prize in computer science is now named after Turing in honor of his great service. The Z3 computer, invented by the German engineer Konrad

Zuse, can write programs. This invention made the world all set for the evolution of the modern computer.

The most commonly thought of computers are desktops and laptops. In fact, the computer also exists in smartphones, cars, home appliances, and other devices. However, no matter how variable the shape, these computers all follow the von Neumann structure. But in the details, there are big differences between computers. Some computers use a multi-level cache, some have only a keyboard without a mouse, and some have tape storage. The hardware of a computer is a very complicated subject. Fortunately, most computer users don't have to deal directly with the hardware. This is due to the operating system (OS).

An operating system is a set of software that runs on a computer and manages its hardware and software resources. Both Microsoft's Windows and Apple's IOS are operating systems. When we program, most of the time it's through the operating system, which is the middleman to deal with the hardware. The operating system provides a set of system calls, which is what the operating system supports. When a system call is called, the computer performs the corresponding operation—alike with pressing a key on a piano, and the piano produces a corresponding rhythm. The operating system, therefore, defines a number of library functions to group system calls to compose a particular function, like a chord made up of several tones—and by programming, we take all these functions and libraries to create beautiful music that is useful.

There are plenty of books on this subject on the market—thanks again for choosing this one! Every effort was made to ensure it is full of as much useful information as possible. Please enjoy!

Chapter 1: Python

This chapter gives a brief introduction to why programming is needed, along with a short introduction about Python.

Why Should You Learn Computer Programming?

Learning any programming language, including Python, opens the door to the computer world. By programming, you can do almost everything a computer can do—giving you plenty of room to be creative. If you think of a need—say, to count the word frequency in a Harry Potter novel—you can program it yourself. If you have a good idea, such as a website for mutual learning, you can open up your computer and start writing. Once you learn to program, you'll find that software is mostly about brains and time, and everything else is extremely cheap. There are many rewards to be gained for writing programs. It could be an economic return, such as a high salary or starting a publicly traded internet company. It can also be a reputational reward, such as making programming software that many people love or overcoming problems that plague the programming community. As hackers and painters put it, programmers are as much creators as painters. Endless opportunities for creativity are one of the great attractions of programming.

Programming is the basic form of human-machine interaction. People use programs to operate machines. Starting with the industrial revolution of the 18th century, people gradually moved away from the mode of production of handicrafts and towards the production of machines. Machines were first used in the cotton industry, and the quality of the yarn produced at the beginning was inferior to that produced by hand. However, the machines can work around the clock, tirelessly, and in prodigious quantities. Hence, by the end of the 18th century, most of the world's cotton had become machine-made. Today, machines are commonplace in our lives. Artificial intelligence is also penetrating more and more into production and life. Workers use machines to make phones and other devices, doctors operate with machines to perform minimally invasive surgery, and traders use machines to trade high-frequency stocks. To put it cruelly, the ability to deploy and possess machines will replace bloodlines and education as the yardstick of class distinctions in the future. *This* is why programming education is becoming more and more important.

Changes in the world of machines are revolutionizing the way that the world works. Repetitive work is dead, and the need for programmers is growing. A lot of people are teaching themselves to program to keep up with the latest trends. Fortunately, programming is getting easier. From assembly language to C to Python, programming languages are becoming more accessible. In Python, for example, a functional implementation requires only a few interface calls and does not require much effort with the support of a rich set of modules. The encapsulation we talked about later is also about packaging the functionality into a canonical interface that makes it easy for others to use. Programming with a precision machine provides the public with a standardized interface. Whether that interface is a fast and secure payment platform or a simple and fast booking site, this encapsulation and interface thinking is reflected in many aspects of social life. Learning to program, therefore, is also a necessary step in understanding contemporary life.

Advantages of Modern Programming

Programming is always calling out the basic instructions of the computer. The code would be incredibly verbose if the entire operation were explained in terms of the basic instructions. In his autobiography, former I.B.M. President John Watson Jr. says he saw an engineer who wanted to do multiplication calculations using punch cards stacked up to 1.2 meters tall. Fortunately, programmers have come to realize that many specific combinations of instructions are repeated. If you can reuse this code in your program, you can save a lot of work. The key to reusing code is called encapsulation.

"Packaging" is the process of packaging an instruction that performs a particular function into a block and giving it a name that is easy to query. If you need to reuse this block, you can simply call it by name. It would be like asking the chef to make a "Chicken Burger" without specifying how much meat, how much seasoning, and how long it takes to cook. The operating system mentioned earlier is used to encapsulate some of the underlying hardware operations for the upper application to call. Of course, encapsulation comes at a cost, as it consumes computer resources. If you're using an early computer, the process of encapsulation and invocation can be time-consuming and ultimately not worth the effort.

There are many ways to encapsulate code. Programmers write programs in a specific style, depending on the way they are written Such as process-oriented programming, object-oriented programming, and functional programming. In more rigorous terms, each style of programming is one Programming Paradigm. Programming languages began to distinguish between camps based on programming paradigms as process-oriented C language, object-oriented Java language, function-oriented Lisp language, etc. A program that is written in any programming paradigm will eventually translate into the simple combination of functions described above. So, programming requirements can always be implemented through a variety of programming paradigms.

Now, the only difference is the convenience of the paradigm. Due to the pros and cons of different paradigms, many modern programming languages support a variety of programming paradigms, allowing programmers to choose between them. Python is a multi-paradigm language.

The programming paradigm is a major obstacle to learning programming. If a programmer is familiar with a programming paradigm, then he can easily learn other programming languages in the same paradigm. You know, for a newbie, learning Python in a multi-paradigm language, you will find different implementation styles for the same function, and you will be puzzled. In some college computer science courses, Cheng, chose to teach each of the typical paradigm languages, such as C, Java, and Lisp, so that students could learn other languages in the future. But doing so can drag on the learning process. It seems to me that a multi-paradigm language like Python provides an opportunity to learn a variety of programming paradigms in contrast. In the same language framework, if the programmer can clearly distinguish between the same programming paradigm, and understand the pros and cons of each respectively, it will cause a multiplier effect. And that's what this book is trying to do—from the aspect, the three main paradigms are the procedure, object-oriented, and functional. Learn Python three times in one book. By learning Python from this book, you will not only learn a Python language, but you'll be able to lay the groundwork for learning other languages in the future.

Why Did Python Emerge as a Winner Among a Lot of Other Computer Languages?

The key to high-level languages is encapsulation, which makes programming easy. Python became one of the major programming languages because it was good at this. Python is widely used and is Google's third-largest development language. It is also the main language used by Dropbox, Quora, Pinterest, Reddit, and other sites. In many scientific fields, such as mathematics, artificial intelligence, bioinformatics, and astrophysics, Python is gaining ground and has a good contribution from programmers all around the globe in GitHub.

Of course, Python's success didn't happen overnight. It has experienced two or three decades of development since its birth. Looking back at the history of Python, we can not only understand the history of Python but also understand the philosophy of it.

What Is Python (and a Little Bit of History)

Python was written by Guido van Rossum, who was Dutch. In 1982, he received a master's degree in mathematics and computer science from the University of Amsterdam. However, although he was a mathematician, he enjoyed the fun of computers even more. In his words, despite his dual aptitude for mathematics and computers, he has always intended to do computer-related work and is keen to do any programming related work.

Before writing Python, Rossum was exposed to and used languages such as Pascal, C, and Fortran. The focus of these languages is to make programs run faster. In the 1980s, though IBM and Apple had already started a wave of personal computers, the configuration of these personal computers seemed to be very low. Early macs had only 8 MHZ OF CPU power and 128 kb of memory, and a slightly more complex operation could cause a computer to crash. Therefore, the core of programming at that time was optimization, so that the program can run smoothly under the limited hardware performance. To be more efficient, programmers have to think like computers so that they can write programs that are more machine-friendly. They want to squeeze every bit of computer power out of their hands. Some even argue that C pointers are a waste of memory. As for the high-level features we now routinely use in programming, such as dynamic typing, automatic memory management, object-orientation, and so on, in those days' computers would simply crash.

However, Rossum fretted about programming with performance as its sole focus. Even if he had a clear idea in his head of how to implement a function in C, the whole writing process would still take a lot of time. Rossum preferred Shell to the C language. Unix system administrators often used the Shell to write simple scripts to do some system maintenance work, such as regular backup, file system management, and so on. The Shell can act as a glue that ties together many of the features under UNIX. Many programs with hundreds of lines in C can be done in just a few lines in the Shell. However, the essence of the Shell is to invoke commands, and it is not a real language. For example, Shell has a single data type and complex operations. In short, the Shell is not a good general-purpose programming language.

Rossum wanted a general-purpose programming language that could call all the functional interfaces of a computer like C and program as easily as a Shell. The first thing that gave Rossum hope was the ABC language. The ABC language was developed by the INFORMATICA in the Netherlands. This institute was where Rossum worked, so Rossum was involved in the development of the ABC language. The purpose of the ABC language was teaching. Unlike most languages of the time, the goal of ABC was "to make users feel better." ABC language hopes to make the language easy to read, easy to use, easy to remember, and easy to learn, in order to stimulate people's interest in learning programming.

Despite the readability and ease of use, the ABC language did not catch on. At the time, the ABC language compiler required a relatively high-profile computer to run. In those days, high-powered computers were a rarity, and their users tended to be computer savvy already. These people were more concerned with the efficiency of the program than with the difficulty of learning the language. In addition to performance, the design of the ABC language suffered from a number of fatal problems:

- Poor extensibility. The ABC language is not a modular language. If you want to add features to the ABC language, such as graphical support, you have to change a lot of things.

- No direct input/output. The ABC language cannot directly manipulate the file system. Although you can import data, such as a text stream, the ABC language cannot read or write files directly. The difficulty of input and output is fatal to computer languages. Can you imagine a sports car that can't open its doors?

- Excessive innovation. The ABC language uses natural language to express the meaning of a program, such as a how-to in the program above. However, for programmers with multiple languages, they are more likely to use a function or to define a function. Similarly, programmers are used to assigning variables with equal signs (=). Innovation, while it makes the ABC language special, actually makes it harder for programmers to learn.

- Transmission difficulty. The ABC compiler is large and must be saved on tape. When Rossum was communicating academically, he had to use a large tape to install the ABC compiler for someone else. This makes it difficult for the ABC language to spread quickly.

In 1989, to pass the Christmas holiday, Rossum began writing a compiler interpreter for Python. The word Python means Python in English. But Rossum chose the name not because of the boa constrictor, but because of a beloved television series. He hopes the new language, called Python, will fulfill his vision. It's a full-featured, easy-to-learn, easy-to-use, and extensible language between C and Shell. As a language design enthusiast, Rossum experimented with design languages. The language design didn't work the last time, but Rossum enjoyed it. This time he designed the Python language, too.

Chapter 2: Importance of Python

In 1991, the first Python compiler or interpreter was born. It is implemented in C language and can call the dynamic link library generated by C language. From the time it was born, Python had the basic syntax it still has today: classes, functions, exceptions, core data types including lists and dictionaries, and a module-based extension system.

What Can You Do as a Python Programmer?

Much of the Python syntax comes from C but is strongly influenced by the ABC language. Like the ABC language, Python uses indentation instead of curly braces, for example, to make the program more readable. According to Rossum, programmers spend far more time reading code than writing it. Forced indenting makes your code easier to read and should be retained. But unlike the ABC language, Rossum also values practicality. While ensuring readability, Python deftly obeys some grammatical conventions that already exist in other languages. Python uses equal sign assignment, which is consistent with most languages. It uses def to define functions, rather than the Esoteric How-to of the ABC language. Rossum argues that there is no need to be overly innovative if it is established by "common sense."

Python also took a special interest in extensibility, which is another embodiment of Rossum's pragmatic principle. Python can be extended at many levels. At a high level, you can extend the functionality of your code by importing Python files written by others. You can also directly import C and C++ compiled libraries for performance reasons. Thanks to years of coding in C and C++, Python stands on the shoulders of a giant. Python is like building a house out of steel, with a large framework and a modular system to give programmers free rein.

The original Python was entirely developed by Rossum himself. Because Python hides many machine-level details and highlights logical level programming thinking, this easy to use language was welcomed by Rossum's colleagues. These colleagues, many of whom were involved in improving the language, were happy to use Python at work and then gave Rossum feedback on how to use it. Rossum and his colleagues made up the core Python team and devoted most of their spare time to Python. Python also gradually spread from Rossum's circle of colleagues to other scientific institutions, slowly used outside the academic community for program development.

The popularity of Python has been linked to significant improvements in computer performance. In the early 1990s, personal computers were introduced into a normal family. Intel released the 486 processor, which represents the fourth generation of processors. In 1993, Intel launched a better Pentium processor. The performance of the computer has been greatly improved. Programmers don't have to work so hard to make their programs more efficient. More and more attention is paid to the ease of use of computers. Microsoft has launched Windows 3.0 with a series of Windows systems that are easy to use graphically. The interface attracted a large number of regular users. Languages that can quickly produce software, such as those running on virtual machines, are the new stars, like Java on a windows phone. Java is entirely based on an object-oriented programming paradigm that can increase program productivity at the expense of performance. Python is a step behind Java, but its ease of use is also up to date. As I said earlier, the ABC language is a failure. One important reason is the performance limitations of the hardware. In this respect, Python is much luckier than the ABC language.

Another quiet change is the Internet. In the 1990s, during the era of the personal computer, Microsoft and Intel dominated the PC market, almost monopolizing it. At the time, the information revolution hadn't arrived, but for coders close to home, the Internet was the tool of the day. Programmers are the first to use the Internet for communication, such as e-mail and newsgroups. The Internet has made it much cheaper to exchange information, and it has allowed people with similar interests to come together across geographic boundaries. Based on the communications capabilities of the Internet, Open Source software development models have become popular. Programmers spend their spare time developing software and opening up source code. In 1991, Linus Torvalds's release of the Linux kernel source on the

Minix newsgroup attracted a large number of coders to join the development effort and led the open-source movement. Linux and GNU work together to form a vibrant open-source platform.

Rossum, himself an open-source pioneer, maintains a mailing list and places early Python uses inside. Early Python users are able to communicate in groups via email. Most of these users are programmers, and they have pretty good development skills. They come from many fields, have different backgrounds, and have various functional requirements for Python.

Because Python is so open and easy to extend, it's easy for a person not to be satisfied with the existing functionality to extend or transform. These users then send the changes to Rossum, who decide whether to add the new feature to Python China. It would be a great honor if the code could be adopted. Rossum's own role is increasingly framed.

If the problem is too complex, Rossum will cut the corner and leave it to the community. Let someone else handle it. Even things like creating websites and raising funds are taken care of. The community is maturing, and the development of the work is divided up among the whole community.

One idea for Python is to have a battery included. That said, Python already has functional modules. The so-called module is that someone else who has written a good Python program can achieve certain functions. A programmer does not need to build wheels repeatedly, just refer to existing modules. These modules include both Python's own standard library and third-party libraries outside of the standard library. These "batteries" are also a contribution from the entire community. Python's developers come from different worlds, and they bring the benefits of different worlds to Python and the regular expression reference in the Python Standard Library.

The syntax for functional programming refers to the LISP language, both of which are contributions from the community. Python provides a rich arsenal of weapons within a concise syntax framework. Whether it's building a website, creating an artificial intelligence program, or manipulating a wearable device, it can be done with an existing library and short code. This is probably the happiest place for a Python programmer.

Of course, Python has its share of pains. The latest version of Python is 3, but Python 3 and Python 2 are incompatible. Since much of the existing code is written in Python 2, the transition from version 2 to version 3 is not easy. Many people have chosen to continue using Python 2. Some people joke that Python 2's version number will increase to in the future 2.7.31415926. In addition to the issue of versioning, Python's performance has been criticized from time to time. Python has low performance in C and C++, and we'll talk about why in this book. While Python is improving its own performance, the performance gap will always be there. But judging from Python's history, a similar critique would be nitpicking. Python itself is a trade performance for usability in the opposite direction of C and C++. Just as a football striker is not a good goalkeeper, and it doesn't make much sense.

For starters, learning to program from Python has many benefits, such as the simple syntax and rich modules mentioned above. Many foreign universities of computer introduction courses have begun to choose Python as the course language, replacing the common use of C or Java. But it is a fantasy to think of Python as the "best language" and to want to learn Python to be the "enemy of all." Every language has its good points, but it also has all sorts of flaws. A language has "good or bad" judgment but is also subject to the platform, hardware, times, and other external reasons. Furthermore, many development efforts require specific languages, such as writing android apps in Java and Apple apps in objective-C or Swift. No matter what language you start with, you will not end up with the language you are just learning. It is only by gambling that the creativity of programming is allowed to flourish.

Example Program to Show You a Good Overview of the Python Programming

Python is easy to install, and you can refer to the next chapter. There are two ways to run Python. If you want to try a few programs and see the results immediately, you can run Python from the command line. The command line is a small input field waiting for you to type on your keyboard and speak directly to Python.

Start the command line, as shown in the next chapter, and you're in Python. Typically, the command line will have a text prompt to remind you to type after the prompt. The Python statements you type are translated into computer instructions by the Python interpreter. We now perform a simple operation: let the computer screen display a line of words. Type the following at the command-line prompt and press Enter on your keyboard to confirm:

>>>print ("Oh, my name is Jesus Christ")

As you can see, when you click enter on the keyboard, the screen then displays:

Oh, my name is Jesus Christ

The input print is the name of a function. The print() function has a specific function, and the print() function simply prints characters on the screen. The function is followed by a parenthesis that says the character you want to print is, "Oh, my name is Jesus Christ." The double quotation marks in parentheses are not printed on the screen. The double quotation marks mark out ordinary characters from program text such as print to avoid confusion on the computer. You can also replace double quotes with a pair of single quotes.

The second way to use Python is to write a Program File that the Python Program file is written in. Py is the suffix, which can be created and written with any text editor. The appendix describes the common text editor for different operating systems. Create a file introduction.py, write the following, and save it:

print("Oh, my name is Jesus Christ")

As you can see, the program content here is exactly the same as it was on the command line. Compared with the command line, program files are suitable for writing and saving a large number of programs.

Run the introduction.py, and you'll see that Python also prints, "Hello World! On the screen!" The contents of the program file are the same as those typed on the command line and produce the same results. Program files are easier to save and change than programs that are entered directly from the command line, so they are often used to write large numbers of programs.

Another benefit of program file is the ability to add comments. Comments are words that explain a program and make it easier for other programmers to understand it. Therefore, the content of the comment is not executed as a program. In a Python program file, where every line starting with # is a comment, we can annotate the hello.py.

print("Oh, my name is Jesus Christ") #Display those words on the screen

If you have too many comments to fit in a single line, you can use a multiline comment.

""" Author: sample Function: Use this to display words """
print('Oh, my name is Jesus Christ')

The multi-line comment glyphs are three consecutive double quotation marks. Multi-line comments can also use three consecutive single quotation marks. Between the two sets of quotation marks is the content of the multi-line comment.

Either the character you want to print or the text you want to annotate can be in any other foreign language. If you use foreign languages in Python 2, you need to add a line of encoding before the program begins to show that the program file uses a utf-8 encoding that supports hundreds of languages. This line of information is not required in Python 3.

- *- coding: utf-8 - *-

So, we wrote a very simple Python program. Don't underestimate the program. In the process of implementing this program, your computer does complicated work. It reads the program file, allocates space in memory, performs many operations and controls, and finally controls the original screen display to display a string of characters. The smooth running of this program shows that the computer hardware, operating system, and language compiler have been installed and set up. So, the first task a programmer does is usually to print a bunch of text on the screen. Meet the Python World for the first time and boom.

Python Reserved Words

Python has certain keywords that cannot be used by any variables because they are distinguished to be used by Python significantly. There are a total number of 33 reserved keywords—and, as, assert, break, def, and delete are a few of the reserved keywords in Python 3.

Chapter 3: How to Install Python

This chapter deals with how to install Python in different operating systems with clear-cut instructions so that newbies will not face any issues while installing Python for the first time.

Installation and Running of Python

Official Version Installation

1) Mac

Python is already pre-installed on the Mac and can be used directly. If you want to use other versions of Python, we recommend using the Homebrew installation. Open the Terminal and enter the following command at a command-line prompt, which brings you to Python's Interactive Command Line:

$python

The Python input above is usually a soft link to a version of a Python command, such as version 3.5. If the corresponding version is already installed, you can run it in the following manner:

$python3.5

The terminal will display information about Python, such as its version number, followed by a command-line prompt for Python. If you want to exit Python, type:

>>>exit()

If you want to run a Python Program in your current directory, append the name to Python or Python 3:

$python installation.py

If the file is not in the current directory, you need to specify the full path of the file, such as:

$python /home/authorname/installation.py

We can also change the installation.py to an executable script. Just add the Python interpreter you want to use to the first line of the installation.py:

#!/usr/bin/env python

In the terminal, change the installation.py to executable:

$chmod installation.py

Then, on the command line, type the name of the program file, and you're ready to run it using the specified interpreter:

$./installation.py

If the installation.py is in the default path, then the system can automatically search for the executable and run it in any path:

$installation.py

2) Linux

Linux systems are similar to MAC systems, and most come preloaded with Python. There are many Linux systems that offer something like Homebrew's software manager, which, for example, is installed under Ubuntu using the following command:

$sudo apt-get install python

Under Linux, Python is used and run in much the same way as on the MAC, and I won't go into that again.

3) Windows Operating System

For the Windows operating system, you need to download the installation package from the official Python Web site. If you don't have access to Python's web site, search engines for keywords like "Python Windows download" to find other download sources. The installation process is similar to installing other Windows software. In the install screen, Customize the installation by selecting Customize, in addition to selecting Python components, check:

Add python.exe to Path

Once installed, you can open the Windows command line and use Python as you would on a Mac.

Other Python Versions

The official version of Python mainly provides compiler / interpreter functionality. Other unofficial versions have richer features and interfaces, such as a more user-friendly graphical interface, a text editor for Python, or an easier to use module management system for you to find a variety of extension modules. In unofficial Python, the two most commonly used are:

1) Anaconda

2) Thought Python Distribution (EPD)

Both versions are easier to install and use than the official version of Python. With the help of a module management system, programmers can also avoid annoying problems with module installation. So, it's highly recommended for beginners. Anaconda is free, and EPD is free for students and researchers. Because of the graphical interface provided, their use method is also quite intuitive. I strongly recommend that beginners choose one of these two versions to use. The exact usage can be found in the official documentation and will not be repeated here.

Virtualenv

You can install multiple versions of Python on a single computer, and using virtualenv creates a virtual environment for each version of Python. Here's how to install virtualenv using Python's included pip.

$pip install virtualenv

You can create a virtual space for a version of Python on your computer, such as:

$virtualenv –p /usr/bin/python3.5 virtualpythonexample

In the above command, /usr/bin/Python3.5 is where the interpreter is located, and virtualPythonexample is the name of the newly created virtual environment. The following command can start using the MYENV virtual environment:

$source virtualpythonexample/bin/activate

To exit the virtual environment, use the following command:

$deactivate

Chapter 4: The World of Variables

This chapter will explain in detail about variables, which are essential for any programming language and are very basic and important for the further understanding of computer languages. This chapter explains variables in detail, along with a lot of examples. Kindly follow the book while using your computer.

The data that appears in few programs—whether it's a number like 1 and 5.2, or a Boolean value like True and False—will disappear after the operation due to various reasons of computer architecture, which is out of the scope of our book. Sometimes, we want to store data in storage to reuse in later programs. Each storage unit in computer memory has an address, like a house number. We can store the data in a cubicle at a particular house number, and then extract the previously stored data from the house number.

But indexing a stored address with a memory address is not convenient because:

- Memory addresses are verbose and hard to remember.

- Each address corresponds to a fixed storage space size; it is difficult to adapt to the type of variable data.

- Before operating on an address, it is not known whether the storage space for that address is already occupied.

With the development of programming language, it began to use variables to store data. Variables, like memory addresses, also serve the function of indexing data. When a variable is created, the computer creates storage space in free memory to store the data, unlike memory addresses, the amount of storage allocated varies depending on the type of variable. The program gives the variable a variable name, which serves as an index to the variable space in the program. The data is given to the variable, and the data is extracted by the name of the variable as needed.

Consider this example:

```
exvariable = "Harry"
 print(exvariable)
#prints out Harry.
```

The output is below:

Harry

In the above program, we pass the value (for example, 276) to the variable exvariable, a process called Assignment. The assignment is represented by an equal sign in Python. With the assignment, a variable is created. From a hardware perspective, the process of assigning a value to a variable is the process of storing the data in memory. A variable is like a small room in which data can be stored. Its name is the house number. The assignment is to send a guest to the room.

"Send Harry to Room exvariable"

In the subsequent addition operation, we extract the data contained by the variable name exvariable and print it out with print(). A variable name is a clue to the corresponding data in memory. With the memory function, the computer won't suffer from Amnesia.

Question: "Who's in the exvariable room? "

Answer: "It's Harry"

Hotel rooms will have different guests check in or check out. The same is true of variables. We can assign one variable to another Value. This changes the guest in the room.

exvariable = "Sachin"
print(exvariable)

prints out "Sachin"

Output is:

Sachin

In computer programming, many variables are often set, with each variable storing data for a different function. For example, a computer game might record the number of different resources that a player has, and it would be possible to record different resources with different variables.

Diamond = 50 # means 50 points
Bronze = 20 # means 20 points
Fire   = 10 # means 10 points

During the game, you can add or subtract resources, depending on the situation. For example, if a player chooses to cut bronze, he adds twenty more points. At this point, you can add 5 to the corresponding variable.

bronze = bronze + 20
print(bronze)
#prints 40

Output is:

40

The computer first performs the operation to the right of the assignment symbol. The original value of the variable is added to 20 and assigned to the same variable. Throughout the game, the variable bronze plays a role in tracking data. The player's resource data is stored properly.

Variable names are directly involved in computation, which is the first step towards abstract thinking. In mathematics, the substitution of symbols for numbers is called Algebra. Many middle school students today will set out Algebraic equations to solve such mathematical problems as "chicken and rabbit in the same cage." But in ancient times, Algebra was quite advanced mathematics. The Europeans learned advanced Algebra from the Arabs, using the symbolic system of Algebra to free themselves from the shackles of specific numbers and to focus more on the relationship between logic and symbols. The development of modern mathematics on the basis of

Algebra laid a foundation for the explosion of modern science and technology. Variables also give programming a higher level of abstraction.

The symbolic representation provided by variables is the first step in implementing code reuse. For example, the code used to calculate the amount of cash needed to buy a house:

50000*(0.3 + 0.1)

When we look at more than one suite at a time, the price of 50000 changes. For convenience, we can write the program as:

Sellingprice= 50000
needed= total *(0.3+ 0.1)
print(needed)
#prints the output

Output is:

20000

This way, you only need to change the value of 50000 each time you use the program. Of course, we'll see more ways to reuse code in the future. But variables, which use abstract symbols instead of concrete numbers, are representative.

Chapter 5: Data Types in Python

This chapter deals with datatypes that variables use. At the start, variables had no data types. However, for the convenience of programmers, data types have been implemented in different programming languages. Python supports a lot of basic data types along with few advanced data types like tuple and dictionary. We will have a detailed description of data types.

Basic Data Types

There may be many different types of data, such as integers like 53, floating-point numbers like 16.3, Boolean values like True and False, and the string "Hello World! ". In Python, we can assign various types of data to the same variable. For example:

datatype_integer = 32
print(datatype_integer)
datatype_string = "My name is Jesus"
print(datatype_string)

As you can see, the value later assigned to the variable replaces the original value of the variable. Python having the freedom to change the characteristics of a variable type is called Dynamic Typing. Not all languages support dynamic typing. In a Static Typing language, variables have pre-specified types. A specific type of data must be stored in a specific type of variable. Dynamic typing is more flexible and convenient than static typing.

Even if you can change it freely, Python's variables themselves have types. We can use the function type() to see the type of the variable. For example:

datatype_integer = 64.34
print(type(datatype_integer))

The output is:

<class 'float'>

float is shorthand for Integer. In addition, there will be numbers (Int), strings (String, str), and Boolean (Boolean, bool). These are the common data types included in the Python programming language.

Computers need to store different types in different ways. Integers can be represented directly as binary numbers, while floating-point numbers have an extra record of the decimal point's position. The storage space required for each type of data is also different. The computer's storage space is in bits (bit) is a unit, each bit can store a 0 or 1 number. To record a Boolean value, we simply have 1 representing the true value and 0 representing the false value. So the store of Boolean values is only 1 bit. For the Integer 4, convert to binary 100. In order to store it, there should be at least three bits of storage, one, zero, and zero, respectively.

For efficiency and utility, a computer must have type storage in memory. In a statically typed language, a new variable must specify a type, and so it is. A dynamically typed language does not need to specify the type but leaves the task of distinguishing the type to the interpreter. When we change the value of a variable, the Python interpreter works hard to automatically identify the type of new data and allocate memory space for that data. Python interpreter's thoughtful service makes programming easier, but it also puts some of your computer's capabilities to work with dynamic types. That's one reason Python isn't as fast as statically typed languages like C.

Tuples and Lists

Some types of variables in Python can hold more than one piece of data as a container. The sequence described in this section, both (Sequence), and the Dictionary in the next section, are container-type variables. Let's start with the sequence. Like a line of soldiers, a sequence is an ordered collection of data. A sequence contains a piece of data called an element of the sequence (element). A sequence can contain one or more elements, or it can be an empty sequence with no elements at all.

There are two types of sequences, Tuples and lists. The main difference between the two is that once established, the elements of a tuple cannot be changed anymore, while the list elements can be changed. So, tuples look like a special kind of table with fixed data.

Therefore, some translators also refer to tuples as "set tables." Tuples and tables are created as follows:

>>>tuplethisis = (65, 9.87, "church", 6.8, 7, True)
>>>listthisis= [False, 4, "laugh"]
>>>type(tuplethisis) # to get the output
>>>type(listthisis) # To get the output

As you can see, the same sequence can contain different types of elements, which is also an embodiment of the Python dynamic type. Also, the element of a sequence can be not only a primitive type of data but also another sequence.

>>>nestlistthisis = [63,[43,84,35]]

Since tuples cannot change data, an empty tuple is rarely created. Sequences can add and modify elements, so Python programs often create empty tables:

>>>thisisempty = []

Since sequences are also used to store data, we cannot help but read the data in the sequence. The elements in a sequence are ordered, so you can find the corresponding elements based on their position in each element. The positional index of a sequence element is called the subscript (Index). The subscript for a sequence in Python starts at 0, which is the corresponding subscript for the first element. There are historical reasons for this, in keeping with the Classic C language. We try to reference elements in the sequence:

The data in the table can be changed so that individual elements can be assigned values. You can use the subscript to indicate which element you want to target.

Dictionaries

In Python, there are also special data types called dictionaries. We will now explain these data types in detail.

Dictionaries are similar to tables in many ways. It is also a container that can hold multiple elements. But dictionaries are not indexed by location. The dictionary allows you to index data in a custom way.

>>>thisisdictionary = {"harry":22, "Ron":37,"Hermoine":64}
>>>type(thisisdictionary) # To input the dictionary values

The dictionary contains multiple elements, each separated by a comma. The dictionary element consists of two parts—a key and a value. The key is the index of the data, and the value is the data itself. The key corresponds to the value one by one. For example, in the example above, "Harry" corresponds to 22, "Ron" corresponds to 37, and "Hermoine" corresponds to 64. Because of the one-to-one correspondence between key values, dictionary elements can be referenced by keys.

>>>thisisdictionary["Ron"]

Output is:

37

Modify or add an element's value to the dictionary:

>>>thisisdictionary["hermoine"] = 57

>>>thisisdictonary["neiville"] = 75

>>>thisisdictionary   # Now the modified is {"Harry": 22, "Ron": 37, "Hermoine": 57, "neiville": 75}

Chapter 6: Operators in Python

Since it is called a "computer," mathematical calculation, of course, is the basic computer skill. The operations in Python are simple and intuitive. Open up the Python Command Line, type in the following numeric operation, and you're ready to run it:

Mathematical Operators

1) Addition

>>>4 + 2

2) Subtraction

>>>4 - 2

3) Multiplication

>>>4 * 2

4) Division

>>>4 / 2

5) Remainder

>>>4 % 2

With these basic operations, we can use Python as if we were using a calculator. Take buying a house. A property costs 20000 dollars and is subject to a 5% tax on the purchase, plus a 10% down payment to the bank. Then, we can use the following code to calculate the amount of cash to be prepared:

>>>20000*(0.5+ 0.1)

In addition to the usual numeric operations, strings can also be added. The effect is to concatenate two strings into one character.

String

Input:

>>>" I am a follower of " + "Christianity"

Output:

I am a follower of Christianity

Input:

>>>"Example" *2

Output:

ExampleExample

Multiplying a string by an integer n repeats the string n times.

Comparison Operator

Python uses comparison operators like ==, >, and < in its program. Below, we will explain with an example.

Program code is below:

```
first = 34
second = 44

if ( first > second)
print  "First one is larger"
else
print "Second one is larger"
```

Output is:

Second one is larger

Logical Operators

In addition to numerical operations, computers can also perform logical operations. It's easy to understand the logic if you've played a killing game or enjoyed a detective story. Like Sherlock Holmes, we use logic to determine whether a statement is true or false. A hypothetical statement is called a proposition, such as "player a is a killer." The task of logic is to find out whether a proposition is true or false.

Computers use the binary system, where they record data in zeros and ones. There are technical reasons why computers use binary. Many of the components that make up a computer can only represent two states, such as the on and off of a circuit, or the high and low voltages. The resulting system is also relatively stable. If you use the decimal system, some computer components will have 10 states, such as the voltage into 10 files. That way, the system becomes complex and error-prone. In Binary Systems, 1 and 0 can be used to represent the true and false states. In Python, we use the keywords True and False to indicate True and False. Data such as True and False are called Boolean values.

Sometimes, we need further logical operations to determine whether a complex proposition is true or false. For example, in the first round, I learned that "player a is not a killer" is true, and in the second round, I learned that "player B is not a killer" is true. So, in the third round, if someone says, "player a is not a killer, and player B is not a killer," then that person is telling the truth. If the two propositions connected by "and" are respectively true, then the whole proposition is true. Virtually, we have a "and" of the logical operation.

In the and operation, when both subpropositions must be true, the compound proposition connected by and is true. The and operation is like two bridges in a row. You must have both bridges open to cross the river, as shown in figure 2-1. Take, for example, the proposition that China is in Asia and Britain is in Asia. The proposition that Britain is in Asia is false, so the whole proposition is false. In Python, we use and for the logical operation of and.

```
>>>True and True  # True
>>>False and True  # false
>>>False and False  # True
```

We can also compound the two propositions with "or." Or is humbler than an aggressive "and." In the phrase "China is in Asia, or Britain is in Asia," for example, the speaker leaves himself room. Since the first half of this sentence is true, the whole proposition is true. "Or" corresponds to the "or" logic operation.

In the "or" operation, as long as there is a proposition for true, then "or" connected to the compound proposition is true. The or operation is like two bridges crossing the river in parallel. If either bridge is clear, pedestrians can cross the river.

The above logic operation seems to be just some life experience and does not need a computer such as complex tools. With the addition of a judgment expression, a logical operation can really show its power.

Operator Precedence

If there is more than one operator in an expression, consider the precedence of the operation. Different operators have different precedence. Operators can be grouped in order of precedence. Below is the list of operator precedence in an order.

Exponent powers have the highest precedence, followed by the mathematical operator multiplication, division, addition, and subtraction. And the next comes Bitwise operators followed by logical operators at the end.

Chapter 7: Execution and Repetitive Tasks

This chapter deals with if-else statement and different types of loop structures that help in repetitive tasks in programming.

If Structure

So far, the Python programs we've seen have been instruction-based. In a program, computer instructions are executed sequentially. Instructions cannot be skipped, nor repeated backward. The first programs were all like this. For example, to make a light come on ten times, repeat ten lines of instructions to make the light come on.

In order to make the program flexible, early programming languages added the function of "jump." With jump instructions, we can jump to any line in the program during execution and continue down. For example, to repeat execution, jump to a line that has already been executed. Programmers frequently jump forward and backward in their programs, for convenience. As a result, the program runs in a sequence that looks like a tangle of noodles, hard to read and prone to error.

Programmers have come to realize that the main function of a jump is to execute a program selectively or repeatedly. Computer experts have also argued that with the grammatical results of "selection" and "loop," "jump" is no longer necessary. Both structures change the flow of program execution and the order in which instructions are executed. Programming languages have entered a structured age. Compared with the "Spaghetti Program" brought about by the "jump," the structured program becomes pleasing to the eye. In modern programming languages, the "jump" syntax has been completely abolished.

Let's start with a simple example of a choice structure: if a house sells for more than $200,000, the transaction rate 3% or 4%. We write a program using a selection structure.

Below is the program code:

price = 340000

```
if
expectedprice> 200000:
fixedtax= 0.04
Else:
fixedtax= 0.03
print(fixedtax)
#prints 0.03
```

Output is:

0.04

In this process, there's an if that we haven't seen before. In fact, the function of this sentence is easy to understand. If the total price exceeds 200,000, then the expected tax is 4%: otherwise, the transaction rate is 3%. The keywords, if and else, each have a line of code attached to them, with a four-space indentation at the beginning of the dependent code. The program will eventually choose whether to execute the if dependent code or the else dependent code, depending on whether the condition after the if holds. In short, the if structure branches off from the program.

If and else can be followed by more than one line:

price = 340000

```
if
total > 200000:
Print ("rate is above 200,000")
fixedtax= 0.04
else:
Print ("rate is below 200,000")
fixedtax= 0.03
print(fixedtax)
# The result is 0.04
```

Output is:

rate is above 200,000

As you can see, code that is also an if or else has four spaces indented. Keywords if and else are like two bosses, standing at the head of the line. There's a little brother standing back from the boss. The boss only by the terms of winning, standing behind his younger brother, has a chance to appear. The last line of the print statement also stands at the beginning of the line, indicating that it is on an equal footing with if and else. The program doesn't need conditional judgment; it always executes this sentence.

Else is not necessary; we can only write if program. For example:

price= 340000
if
price > 200000:
Print ("total price over $200,000")

Without else, it is effectively equivalent to an empty else. If the condition after if doesn't hold, then the computer doesn't have to do anything.

Stand Back

Python features indentation to indicate the dependencies of your code. As we show, the design for indenting code relationships is derived from the ABC language. For comparison, let's look at how C is written:

if (price > 0) {
selling= 1;
buying= 2;
}

The program means that if the variable price is greater

than 0, we will do the two assignments included in the parentheses. In C, a curly brace is used to represent a block of code that is subordinate to the if. The average programmer also adds indentation to the C language to distinguish the dependencies of instructions. But indenting is not mandatory. The following does not indent the code, in the C language can also be normal execution, and above the results of the program run no difference:

In Python, the same program must be written in the following form:

```
if price> 0:
    selling= 1
    buying= 2
```

In Python, the parentheses around 0 are removed, the semicolon at the end of each statement is removed, and the curly brackets around the block are also removed. There's more—the colon (:) and the indentation of four spaces in front of 1 and 2. By indenting, Python recognizes that both statements are subordinate to the if. Indenting in Python is mandatory in order to distinguish between dependencies. The following procedure will have an entirely different effect:

```
if price> 0:
selling= 1
buying= 2
```

Here, only selling that is 1 is subordinate to if, and the second assignment is no longer subordinate to if. In any case, buying will be assigned to 2.

It should be said that most mainstream languages today, such as C, C++, Java, JavaScript, mark blocks with curly braces, and indentation is not mandatory. This syntax is derived from the popularity of the C language. On the other hand, while indenting is not mandatory, experienced programmers write programs in these languages with indenting in order to make them easier to read. Many editors also have the ability to indent programs automatically. Python's forced indentation may seem counterintuitive, but it's really just a matter of enforcing this convention at the syntactic level so that programs look better and are easier to read. This way of writing, with four spaces indented to indicate affiliation, is also seen in other Python syntax constructs.

If Nesting and Elif

And then back to the choice of structure. The choice of structure frees the program from the tedium of command-and-control permutations. A program can have a branching structure inside of it. Depending on the conditions, the same program can work in a volatile environment. With Elif Syntax and nested use of if, programs can branch in a more colorful way.'

The next program uses the elif structure. Depending on the condition, the program has three branches:

result= 1
if result > 0:# Condition 1. Since I is 1, this part will perform.
print("positive result")
result= result + 1
elif result == 0:

Condition 2. This part is not executed.

print("result is 0")
result= result*10
Else:

```
# Condition 3. This part is not executed.
print("negative result")
result= result - 1
```

There are three blocks, led by if, Elif, and else. Python first detects the condition of the if, skips the block that belongs to the if the condition is false, and executes the else block if the condition of the Elif is still false. The program executes only one of three branches, depending on the condition. Since the result has a value of 1, only the if part is executed in the end. In the same way, you can add more elif between if and else to branch your program.

We can also nest an if structure inside another if structure:

```
result = 5
if  result> 1:    # This condition holds, execute the Internal Code
print("result bigger than 1")
print("nice")

if result > 2:    # nested if structure, the condition holds.
print("result bigger than 2")
print("Its good than before")
```

After making the first if judgment, if the condition holds, the program runs in sequence and encounters the second if construct. The program will continue to judge and decide whether to execute based on the conditions. The second subsequent block is indented four more spaces relative to the if to become the "little brother." Programs that further indent are subordinate to the inner if. In general, with the if construct, we branch the program. Depending on the conditions, the program will take a different path.

For Loop

Loops are used to iterate through blocks of code. In Python, loops are either for or while. Let's start with the for loop. From the selection structure in section 2.3, we have seen how to use indentation to represent the membership of a block. Loops use similar notation. Programs that are part of a loop and need to be repeated are indented, such as:

```
for input in [4,6.8,"love"]:
print(c)   #prints each element in the list in turn
```

The loop is to take an element from the list [4,6,8, "love"] one at a time, assign it to c, and execute the line belongs to the program for, which calls the print()

function to print out the element. As you can see, one of the basic uses of for is to follow in with a sequence:

For Element in Sequence

The number of elements in a sequence determines the number of iterations. There are three elements in the example, so print() will be executed three times. That is, the number of repetitions of the for loop is fixed. The for loop, in turn, takes elements from the sequence and assigns them to the variable immediately after the for (a) in the example above. Therefore, even though the statements executed are the same, the effect of the same statement changes after three executions because the data also changes.

One of the conveniences of a for loop is to take an element from a sequence, assign it to a variable, and use it in a membership program. But sometimes, if we simply want to repeat a certain number of times and don't want to create a sequence, we can use the range() function provided by Python:

```
for result in range(3):
print("This is crazy")   # print "This is crazy" Three Times
```

The 3-way range() function in the program indicates the number of times you want to repeat. As a result, the program that belongs to for is executed five times. Here, after the for loop, there is still the variable I, which counts for each loop:

```
for result in range(7):
print(result, "This is crazy") # prints the sequence number and "This is crazy"
```

As you can see, the count provided by range() in Python also starts at 0, the same as the index of the table. We also saw a new use of print(), which is to specify multiple variables in parentheses, separated by commas. The function print() prints them all.

Let's look at a practical example of a for loop. We have previously used Tuples to record the yearly interest rate on a mortgage.

```
thisisinteresttuple= (0.06, 0.07, 0.08, 0.1, 0.3)
```

If there is a 200,000-dollar mortgage, and the principal is unchanged, then the annual interest to pay is how much? Using the for loop:

```
price= 200000
for interest in thisisinteresttuple:
debt= price* interest
Print ("you need to pay ", debt)
```

While Loop

There is also a loop structure in Python, the while loop. The use of the While is:

```
check= 0
while check< 20:
print(check)
check= check+ 1

# prints from 0 to 19
```

A while is followed by a condition. If the condition is true, the while loop continues to execute the statements that belong to it. The program will only stop if the condition is false. In the while membership, we change the variable I that participates in conditional judgments until it becomes 10 so that the loop is terminated before the condition is met. This is a common practice for the while loop. Otherwise, if the while condition is always true, it becomes an infinite loop.

Once there is an infinite loop, the program will continue to run until the program was interrupted or the computer shuts down. But sometimes, infinite loops can be useful. Many graphics programs have an infinite loop to check the status of the page and so on. If we were to develop an infinite ticket-snatching program, an infinite loop would sound good. The infinite loop can be written in a simple way:

```
while
True:
print("This is crazy")
```

Skip or Abort

The loop structure also provides two useful statements that can be used inside the loop structure to skip or terminate the loop. Continue skips this execution of the loop for the next loop operation. Break stops the whole loop.

```
for result in range(20):
```

```
if result == 2:
Continue
print(result)
# prints 0,1,3,4,5,6,7,8,9,11,13,15,17,19 notice that you skipped 2
```

When the loop executes until the result is 2, the if condition holds, triggering continue, instead of printing result at this point—the program proceeds to the next loop, assigns 3 to result, and continues to execute the for subordinate statement. The continue simply skips a loop, while the break is much more violent, and it terminates the entire loop.

```
for result in range(20):
if result == 2:
Break
Print (result) # prints only 0 and 1
```

When the loop reaches 2, the if condition holds, triggers the break, and the entire loop stops. The program no longer executes the statements inside the for loop.

Small Exercise to Review What We Learned Until Now

In this chapter, we learned about operations and variables, as well as about the selection and circulation of two process control structures. Now, let's do a more complicated exercise and go over what we learned together.

Suppose I could get a full loan to buy a house. The total price of the house is half a million. In order to attract buyers, the first four years of mortgage interest rate discount, respectively 3%, 4%, 5%, 6%. For the rest of the year, the mortgage rate was 5% a year. I pay back the money year after year, up to 200000 $ each time. So, how many years will it take to pay off the house fully?

Think about how you can solve this problem with Python. If you think clearly, you can write a program to try it. The best way to learn to program is to get your hands dirty and try to solve problems. The following is the author's solution, for reference only:

```
price= 0
selling= 200000
thisisinterest= (0.03, 0.04, 0.05, 0.06)
```

```
debt= 10000
while debt > 0:
price= price + 1
Print ("yes", "result", "year or money")
if result<= 4:
interest = thisisinterest[result- 1]
# The subscript of the sequence starts at zero
Else:
interest = 0.05
debt= debt* (debt + 1) - price
Print ("yes", result + 1, "the year is finally over")
# secretly, the 23rd year is over
```

Chapter 8: Functions and Modules

In this chapter, we'll look at other process-oriented encapsulation methods—namely, functions and modules. Functions and modules encapsulate chunked instructions into blocks of code that can be called repeatedly and organize a set of interfaces with function names and module names to facilitate future calls.

What Are Functions?

It's a bit of a pain because, whenever you think of functions, you will remember the topic you have learned in mathematics. In mathematics, a function represents a correspondence between sets. For example, all books are a collection, and all pens are a collection. There is a correspondence between the set of books and the set of pens, which can be expressed as a function.

Let's take one mathematical example. The following

square function maps a natural number to the cubes of the natural number:

$f(x) = x^3$

(where X is a natural number)

In other words, the function f (x) defines the correspondence between two sets of numbers:

```
x -> y
1 1
2 8
3 27
4 64
..
..
```

A mathematical function defines a static correspondence. From a data point of view, a function is like a magic box that transforms a walking pig into a rabbit. For the function f (x) just defined, what goes in is a natural number, and what comes out is the cube of that natural number. With the function, we implement the data transformation.

The magic transformation of a function does not happen out of thin air. For a function in programming, we can use a series of instructions to show how the function works. In the programming function is the realization data transformation, but also may, through the instruction, realize other functions. So, the programmer can also understand functions from the perspective of program encapsulation.

For programmers, a function is such a syntactic construct. It encapsulates a number of commands into a single punch. Once the function is defined, we can start the combination by calling the function. Therefore, a function is an exercise in encapsulation philosophy. The input data is called a parameter, which affects the behavior of the function. It's as if the same combination can have different levels of power.

Thus, we have three ways of looking at functions: The correspondence of sets, the magic box of data, and the encapsulation of statements. Programming textbooks generally choose one of these to describe what a function is. All three explanations are correct. The only difference is the perspective. By cross-referencing the three interchangeable interpretations, you can better understand what a function is.

Defining Functions

Let's first make a function. The process of making a function is also called defining a function. We call this function squareofnum(). As the name suggests, the function calculates the sum of the squares of two numbers:

```
def squareofnum(first,second):

first = first**2
second = second**2
result = first+ second
return result
```

The first keyword to appear was "def". This keyword tells Python, "here comes the definition of the function." The keyword def is followed by squareofnum, the name of the function. After the function name, there are parentheses to indicate which arguments the function takes—namely, first and second—in parentheses. Parameters can be multiple in number or none at all. According to Python Syntax, the parentheses following a function should be preserved even if no input data is available.

In defining the function, we use symbols(variables) first and second to refer to the input data. Until we actually use the function, we won't be able to specify what numbers first and second are. So, defining a function is like practicing martial arts. When you actually call a function, you use the actual input data to determine how hard you want to hit it. Arguments function like variables inside a function definition, participating in any line of instruction in a symbolic form. Because the parameter in a function definition is a formal representation, not real data, it is also called a parameter.

In defining the function squareofnum(), we complete the symbolic square summation with parameters first and second. In the execution of a function, the data represented by the parameter does indeed exist as a variable, as we will elaborate on later.

At the end of the parentheses, you come to the end of the first line. There's a colon at the end, and the last four lines are indented. We can infer that the colon and indenting represent the subordination of the code. So, the four indented lines of code that follow are the kids of the function squareofnum(). A function is an encapsulation of code. When a function is called, Python executes the statements that belong to the function until the end of the statement. For squareofnum(), the first three lines are the familiar arithmetic statements. The last sentence is returned. The keyword return is used to describe the return value of a function, which is the function's output data.

As the last sentence of a function, the function ends on return, regardless of whether there are other function definition statements after it. If you replace squareofnum() with the following:

```
def squareofnum(first,second):

first = first **2
second = second**2
result = first+ second
return result
```

Print (" Here ends the result")

Then, when the function executes, it will only execute to return result. The latter statement, print(), although also a function, is not executed. So, return also has the ability to abort the function and specify the return value. In Python Syntax, return is not required. If there is no return, or if there is no return value after return, the function returns None. None is the empty data in Python, used to indicate nothing. The Return Keyword also returns multiple values. Multiple values are followed by return, separated by Commas.

How to Call Functions?

Above, we see how to define a function. Defining a function is like building a weapon, but you have to use it to make it work. The procedure that uses a Function is called a Call Function. In the previous chapter, we've seen how to call the print() function:

print("Hello python universe!")

We use the function name directly, with specific arguments in parentheses. The argument is no longer the symbol used to define the function, but the actual data string, "Hello Python universe!" Therefore, arguments that occur during a function call are called arguments.

The function print() returns a value of None, so we don't care about the return value. But if a function has another return value, then we can get the return value. A common practice is to assign the return value to a variable for later use. The squareofnum() function is called in the following program:

```
result =squareofnum(4,6)
print(result)
```

Python knows that 4 corresponds to the first parameter in the function definition, 6 corresponds to the second parameter second and passes the parameter to squareofnum(). The function squareofnum() executes the internal statement until it returns the value 52. The return value of 52 is assigned to the variable result, which is printed by print().

Function calls are written as they were written after def, the first line of the function definition. It's just that when we call a function, we put real data in parentheses and pass it as an argument to the function. In addition to specific data expressions, parameters can be variables that already exist in the program, such as:

```
first = 8
second = 9
result = squareofnum(first, second)
print(x)
```

Function Documentation

Functions can encapsulate code and reuse it. For some frequently called programs, if you can write a function and call it every time, it will reduce the amount of repetitive programming. However, too many functions can cause problems. The common problem is that we often forget what a function is supposed to do. Of course, you can find the code that defines the function, read it line by line, and try to understand what you or someone else is trying to do with it. But the process sounds painful. If you want your future self or others to avoid similar pain, you need to write functions with clear documentation of what the functions do and how they are used.

We can use the built-in function help() to find the documentation for a function. Take the function min() for example. Use this function to return the maximum value. For example:

variable= min(6,7,9,12)
print(variable)
# result is 6

The function min() takes multiple arguments and returns the largest of them. If we can't remember the function min() and its parameters, we can ask help().

>>> help(min)

Help on built-in function max in module __builtin__:

min(...)
min(iterable[, key=func]) -> value
min(d, e, f,...[, key=func]) -> value
With a single iterable argument, return its largest item.
With two or more arguments, return the largest argument.
(END)

As you can see, the function Min() is called in two ways. Our previous call was in the second way. In addition, the documentation explains the basic functions of the function Min().

The function Min() belongs to Python's own defined built-in function, so the documentation is ready in advance. For our custom functions, we need to do it ourselves. The process is not complicated, so here's a simple notation for the function cube():

def sumdefined(first,second):

"""return the square sum of two arguments"""

first= first**2
second= second**2
third = first+ second
return third

At the beginning of the function content, a multi-line comment is added. This multi-line comment is also indented. This will be the documentation for the function. If I use the function help() to view the documentation for square(), help() will return what we wrote when we defined the function:

```
>>>help(sumdefined)

Help on function sumdefined in module __main__:

sumdefined(a, b)

return the square sum of two arguments
```

In general, the documentation should be as detailed as possible, especially for the parameters and return values that people care about.

Parameter Passing

Basic Pass Parameters

Passing data as arguments to a function is called parameter passing. If there is only one parameter, then parameter passing is simple, simply mapping the only data entered at the time of the function call to this parameter. If you have more than one parameter, Python determines which parameter the data corresponds to when you call the function based on its location. For example:

```
def argument(first, second, third):
"""print arguments according to their sequence"""
```

```
print(first, second, third)
```

```
print_arguments(1, 3, 5)
print_arguments(5, 3, 1)
print_arguments(3, 5, 1)
```

In each of the three calls to the program, Python determines the relationship between the arguments by their locations. If you find that positional arguments are rigid, you can pass them in the form of keywords. When we define a function, we give the parameter a symbolic token, which is the parameter name. Keyword passing is based on the parameter name to make the data and symbols on the corresponding. Therefore, if the keyword is passed in at the time of the call, the correspondence of the location is not followed. Use The function definition above and pass it as a parameter instead:

```
print_arguments(third=5,second=3,first=1)
```

As you can see from the results, Python no longer uses locations to correspond to parameters but instead uses the names of parameters to correspond to parameters and data. Positional and keyword passing can be used together, with one part of the argument being passed based on location and the other on the name of the argument. When a function is called, all positional arguments appear before the keyword arguments. Therefore, you can call:

print_arguments(1, third=5,second=3)

But if you put the position parameter 1 after the keyword parameter C5, Python will report an error:

print_arguments(third=5, 1, second=3)

Position Passing and keyword passing allow data to correspond to formal parameters, so the number of data and formal parameters should be the same. But when defining a function, we can set default values for certain parameters. If we do not provide specific data for these parameters when we invoke them, they will take the default values defined, such as:

def f(first,second,third=10):
return first+second+third
print(f(3,2,1))
print(f(3,2))

The first time you call the function, you enter three data, which corresponds to three parameters, so the parameter C corresponds to 1. On the second call to the function, we provided only 3 and 2 data. The function maps 3 and 2 to the shape parameters A and B, depending on the position. By the time we get to parameter C, there is no more data, so C will take its default value of 10.

Pass the Parcel

All of the above methods of passing arguments require that the number of arguments is specified when defining the function. But sometimes when we define a function, we don't know the number of arguments. There are many reasons for this, and sometimes it's true that you don't know the number of parameters until the program is running. Sometimes, you want the function to be more loosely defined so that it can be used for different types of calls. At this point, it can be very useful to pass parameters by packing them.

As before, the package pass-through takes the form of a location and a keyword. Here's an example of a package position pass:

def place(*all_arguments):
print(type(all_arguments))

```
print(all_arguments)
postion (2,7,9)
position(5,6,7,1,2,3)
```

Both calls are based on the same package() definition, although the number of parameters is different. Calling package(), all data is collected into a tuple in sequence. Inside the function, we can read incoming data through tuples. That's the package. Pass it on. In order to remind you that the Python parameter all is a package. When we define package(), we prefix the tuple name all with an asterisk.

Let's take another look at the package keyword pass example. This parameter passing method collects the incoming data into a dictionary:

```
def package(**arguments):
print(type(arguments))
print(arguments)
package(first=1,second=9)
package(fourth=2,fifth=1,third=11)
```

Similar to the previous example, when a function is called, all parameters are collected into a data container. However, when the package keyword is passed, the data container is no longer a tuple, but a dictionary. Each parameter call, in the form of a keyword, becomes an element of the dictionary. The parameter name becomes the element's key, and the data becomes the element's value. All the parameters are collected and passed to the function. As a reminder, the parameter all is the dictionary used to package keyword delivery, so you prefix all with * *.

The package location pass parameter and the package keyword pass parameter can also be used together. For example:

```
def package(*place, **keywords):
print(place)
print(keywords)
package(1, 2, 3, a=7, b=8, c=9)
```

You can go a step further and mix the package pass with the basic pass. They appear in the order is Location → Keyword → package location → package keyword. With parcel passing, we have more flexibility in representing data when defining functions.

Unwrap

In addition to being used for function definition, * and * * Can also be used for function calls. In this case, both are to implement a syntax called unpacking. UNWRAPPING allows us to pass a data container to the function and automatically decompose it into arguments. It should be noted that the package transferring and UNWRAPPING is not the opposite operation, but two relatively independent functions. Here's an example of UNWRAPPING:

```
def packagediscontinue(first,second,third):
print(first,second,third)
args = (1233,42)
packagediscontinue(*args)
```

In this example, packagediscontinue() uses the basic pass-through method. The function takes three arguments and passes them by position. But when we call this function, we know about the package. As you can see, we are passing a tuple when we call the function. A tuple cannot correspond to three parameters in the way a primitive argument is passed. But we can do that by prefacing "args" with an asterisk (*).

Remind Python that I want to break a tuple into three elements, each of which corresponds to a positional parameter of the function. Thus, the three elements of a tuple are assigned three parameters.

Accordingly, the dictionary can also be used for Unwrapping, using the same unpackage() definition:

args = {"first":1, "second":2, "third":3}
packagediscontinue(**args)

Then, when passing the dictionary args, each key-value pair of the dictionary is passed as a keyword to the function packagediscontinue().

Unwrapping is used for function calls. When a function is called, several arguments can also be passed in a mix. It's still the same basic principle: Location → Keyword → location Unwrapping → keyword unwrapping.

Recursion

GAUSS and Mathematical Induction

Recursion is the operation of a function call itself. Before we get to recursion, let's take a look at a short story by the mathematician GAUSS. It is said that once the teacher punished the whole class by having to work out the sum of 1 to 100 before going home. Only seven years old, Gauss came up with a clever solution that became known as the gauss summation formula. Here's how we'll solve GAUSS's summation programmatically:

```
addition= 0
for result in range(1, 101):
addition= addition + i
print(addition)
```

As the program shows, a loop is a natural way to solve a problem. But this is not the only solution, we can also solve the problem in the following ways:

```
def sumgauss(result):
if result == 1:
return 1
Else:
```

return result+ sumgauss(result-1)
print(sumgauss(100))

The above solution uses Recursion, in which the function itself is called within a function definition. In order to ensure that the computer does not get stuck in a loop, recursion requires that the program have a Base Case that can be reached. The key to recursion is to show the join condition between the next two steps. For example, we already know the cumulative sum of 1 to 64, which is Gaussian(64), then the summation of 1 to 64 can easily be found: Gaussian(64) Gaussian(63) + 62.

When we use recursive programming, we start with the end result, which is that in order to find Gaussian(100), the computer breaks the calculation down to find Gaussian(99) and to find Gaussian(99) plus 100. And so on, until you break it down into Gaussian(1), then you trigger the termination condition, which is N1 in the if structure, to return a specific number 1. Although the whole process of recursion is complicated, when we write a program, we only need to focus on the initial conditions, the termination conditions, and the join, not on the specific steps. The computer will be in charge of the execution.

Recursion comes from mathematical induction. Mathematical Induction is a Mathematical proof, often used to prove that a proposition is valid in the range of natural numbers. With the development of modern mathematics, proofs within the scope of natural numbers have actually formed the basis of many other fields, such as mathematical analysis and number theory, so mathematical induction is of vital importance to the whole mathematical system.

The mathematical induction itself is very simple. If we want to prove a proposition for the natural number N, then:

The first step is to prove that the proposition holds for n 1.
The second step is to prove that the proposition holds for N + 1 under the assumption that N is an arbitrary natural number.

Proof of the Proposition

Think about the two steps above. They actually mean that the proposition holds for N1→ the proposition holds for N2→ the proposition holds for N3, and so on, until infinity. Therefore, the proposition holds for any natural number. It's like a domino. We make sure that N goes down, causes N + 1 to go down, and then we just push down the first domino to make sure that any domino goes down.

Function Stack

Recursion in the program requires the use of the Stack data structure. The so-called data structure is the organization of data stored by a computer. The stack is a kind of data structure, which can store data in an orderly way.

The most prominent feature of the stack is "Lifo, Last In, First Out". When we store a stack of books in a box, the books we store first are at the bottom of the box, and the books we store later are at the top. We have to take the books out of the back so that we can see and take out the books that were in the first place. This is Lifo. The stack is similar to this book box, only "last in, first out". Each book, that is, each element of the stack, is called a frame. The stack supports only two operations: Pop and push. The stack uses pop operations to get the top element of the stack and push operations to store a new element at the top of the stack.

As we said before, in order to Compute Gaussian(100), we need to pause Gaussian(100) and start computing Gaussian(99). To calculate Gaussian(99), pause Gaussian(99), and call Gaussian(98), and so on. There will be many incomplete function calls before the termination condition is triggered. Each time a function is called, we push a new frame into the stack to hold information about the function call. The stack grows until we figure out Gaussian(1), then we go back to Gaussian(2), Gaussian(3), and so on. Because the stack is "backward advanced."

"Out" feature, so each time just pop up the stack frame, it is what we need Gaussian(2), Gaussian(3), and so on until the pop-up hidden in the bottom frame Gaussian(100).

Therefore, the process of running a program can be seen as a stack first growth and then destroy the stacking process. Each function call is accompanied by a frame being pushed onto the stack. If there is a function call inside the function, another frame is added to the stack. When the function returns, the corresponding frame is pushed off the stack. At the end of the program, the stack is cleared, and the program is complete.

Scope of Variables

With a function stack in place, the scope of a variable becomes simple. A new variable can be created inside a function, such as the following:

```
def variable(first, second):
    third= first+ second
    return third
print(variable(4, 6))
```

In fact, Python looks for variables in more than the current frame. It also looks for variables that are defined outside the function in Python's main program. So, inside a function, we can "see" variables that already exist outside the function. For example, here's the program:

```
def variable():
print(result)
result= 5
```

inner_var()

When a variable is already in the main program, the function call can create another variable with the same name by assigning it. The function takes precedence over the variable in its own function frame. In the following program, both the main program and the function external() have an info variable. Inside the function external(), the info inside the function is used first:

```
def externalvariable():
detail = "Authors Python"
print(detal)
detail= "This is crazy"
externalvariable()
print(detail)
```

And the function uses its own internal copy, so the internal action to Info does not affect the external variable info. The arguments to a function are similar to the arguments inside the function. We can think of a parameter as a variable inside a function. When a function call is made, the data is assigned to these variables. When the function returns, the variables associated with these parameters are cleared. But there are exceptions, such as the following:

When we pass a table to a function, table B outside the function changes. When the argument is a data container, there is only one data container inside and outside of the function, so the operation inside the function on the data container affects the outside of the function. This involves a subtle mechanism in Python that we'll explore in more detail. Now, it is important to remember that for the data container, changes inside the function affect the outside.

Introducing Modules

There used to be a popular technical discussion on the Internet: "How do you kill a dragon with a programming language?" There were many interesting answers, such as the Java Language "Get out there, find the Dragon, develop a multi-tier dragon kill framework, and write a few articles about it... but the dragon wasn't killed." The answer was a mockery of Java's complex framework. "C."

The language is: "Get there, ignore the Dragon, raise your sword, cut off the Dragon's head, find the princess... hang the princess." The answer is to praise the power of the C language and the commitment of the C community to the Linux kernel. As for Python, it's simple:

Import   functionlibrary;

People who know Python modules smile at this line of code. In Python, A. The. Py file constitutes a module. With modules, you can call functions in other files. The import module was introduced to reuse existing Python programs in new programs. Python, through modules, allows you to call functions in other files. Let's start with a first.py document that reads as follows:

```
def smile():
print("HaHaHaHa")
```

And write a laugh.py file in the same directory. Introduce the first module into the program:

```
from first import smile
for result in range(10):
smile()
```

With the import statement, we can use the smile() function defined in the laugh.py in the URL. In addition to functions, we can also introduce data contained in other files. Let's say we're in a module (trail.Py). Write:

```
text = "Hello programmer"
```

In import second.Py, we introduce this variable:

```
from import_demo import text
Print (text) # prints 'Hello programmer"
```

For process-oriented languages, a module is a higher-level encapsulation pattern than a function. Programs can be reused in units of files. A typical process-oriented language, such as C language, has a complete module system. The so-called Library consists of common functions programmed into modules for future use. Because Python's libraries are so rich, much of the work can be done by reference libraries, drawing on the work of previous generations. That's why Python uses the import statement to kill dragons.

Search Path

When we introduced the module just now, we put the library file and the application file in the same folder. When you run a program under this folder, Python automatically searches the current folder for modules it wants to introduce.

However, Python also goes to other places to find libraries:

(1) the installation path of the Standard Library
(2) the path contained in the operating system environment variable PYTHONPATH
The Standard Library is an official library of Python.

Python automatically searches the path where the standard library is located. As a result, Python always correctly introduces modules from the Standard Library. For example:

import time

If you are a custom module, put it where you see fit and change the Python path environment. When Python path contains the path of a module, Python finds that module.

When Python introduces a module, it goes to the search path to find the module. If the introduction fails, it is possible that the search path was set incorrectly. We can set the search path as follows.

Inside Python, you can query the search path in the following way:

>>>import best
>>>print(best.path)

As you can see, best.path is a list. Each element in the list is a path that will be searched. You can control the search path for Python by adding or removing elements from this list.

The above change method is dynamic, so each time you write the program, you add related changes. We can also set the PYTHONPATH environment variable to change the Python search path statically. On Linux, you can use them. Add the following line to the bashrc file to change the PYTHONPATH:

export PYTHONPATH=/home/user/mylib:$PYTHONPATH

The meaning of this line is to add /home/user/mylib to the original PYTHONPATH. The files that need to be modified under the MAC are under the home folder. The modification method is similar to Linux.

You can also set a PYTHONPATH on Windows. Right-click your computer and select properties from the menu. A system window appears. Click the advanced system settings, and a window called system properties appears. Select the environment variable, add a new variable to the PYTHONPATH, and set the value of that variable to the path you want to search for.

Installation of Third-Party Modules

In addition to the modules in the standard library, there are many third-party contributed Python modules. The most common way to install these modules is to use PIP. PIP is also installed on your computer when you install Python. If you want to install third-party modules such as Numpy, you can do so in the following manner:

$pip install numpy

If you use VIRTUALENV, each virtual environment provides a corresponding pip to the Python version of the virtual environment. When you use a pip in an environment, the module is installed into that virtual environment. If you switch to virtuality, the modules you use and the versions of the modules you use will change, avoiding the embarrassment of modules not matching the Python version.

Additional tools for installing third-party modules are available under EPD Python and Anaconda and can be found at the official website. You can use the following command to find all installed modules, as well as the version of the module:

$pip freeze

Chapter 9: Reading and Writing Files in Python

Once you understand the basics of object-oriented programming, you can take advantage of the wide variety of objects in Python. These objects can provide a wealth of functionality, as we will see in this chapter for file reading and writing, as well as time and date management. Once we get used to these powerful objects, we can implement a lot of useful features. The fun of programming is to implement these functions in a program and actually run them on a computer.

Storage

Documents

We know that all the data in Python is stored in memory. When a computer loses power, it's like having amnesia, and the data in the memory goes away. On the other hand, if a Python program finishes running, the memory allocated to that program is also emptied. For long-term persistence, Python must store data on disk. That way, even if the power goes out or the program ends, the data will still be there.

The disk does store data in file units. For a computer, the essence of data is an ordered sequence of binary numbers. If you take a sequence of bytes, that is, every eight bits of a binary number, then that sequence of data is called text. This is because an 8-bit sequence of binary numbers corresponds to exactly one character in ASCII encoding. Python, on the other hand, can read and write files with the aid of text objects.

In Python, you can create file objects with the built-in function open. When you call open, you need to specify the file name and how to open the file:

f = open (filename,method)

A filename is the name of a file that exists on disk. Common ways to open a file are:

"r"   # to read an existing file
"w" # create a new file and write
"a" # if the file exists, write to the end of the file. If the file does not exist, a new file is created and written

For example:

>>>f = open("harrypotter.txt","r")

This will instruct Python to read a text file named harrypotter. It's a read-only way to open a file called harrypotter.txt.

With the object returned above, we can read the file:

file= f.read(30)

Read 30 bytes of data

file= f.readline()

read a line

file= f.readlines()

read all the rows and store them in a list, one row for each element.

If it is opened "w" or "a", then we can write text:

f = open("Harry.txt", "w")

```
f.write("This is hogwarts")
```

```
# write " This is hogwarts" to the file
```

If you want to write a line, you need to add a newline at the end of the string. On UNIX systems, the line feed is "\n". In Windows, the line feed is "\r \n".

Example:

```
f.write(" This is the chamber of secrets \n")      # UNIX
```

```
f.write("This is the chamber of secrets \r\n") # Windows
```

Opening the file port takes up computer resources, so close the file in a timely manner with the close method of the file object after reading and writing.

```
f.close()
```

Context Manager

File operations are often used with context managers. The context manager is used in specifying the scope of use for an object. Once in or out of this scope, special actions are called, such as allocating or freeing memory for an object. The context manager can be used for file operations. For file operations, we need to close the file at the end of the read and write. Programmers often forget to close files, taking up resources unnecessarily. The context manager can automatically close files when they are not needed.

Here is an example of file manipulation.

```
f = open("Ron.txt", "w")
 print(f.closed)             # check to see if the file is open
f.write("I love Quidditch")
f.close()
print(f.closed)              # print True
```

If we add the context manager syntax, we can rewrite the program to:

```
# use the context manager

with open("new.txt", "w") as f:
        f.write("Hello World!")

print(f.closed)
```

The second part of the program uses with... as... Structure. The context manager has a block that belongs to it, and when the execution of that block ends, that is, when the statement is no longer indented, the context manager automatically closes the file. In the program, we call F. Closed property to verify that it is closed. With a context manager, we use indentation to express the open range of a file object. For Complex programs, the presence of indenting makes the coder more aware of the stages at which a file is opened, reducing the possibility of forgetting to close the file.

The above context manager is based on the () special method of the f object. When using the syntax of the context manager, Python calls the () method of the file object before entering the block, and the file pair at the end of the Block. The () method of image. In the () method of the file object, there is self. Close () statement. Therefore, we do not have to close the file in clear text when using the context manager.

Any object that defines a () method and a () method can be used by the context manager. Next, we customize a class ram and define its () and () methods. Thus, objects from the Vow class can be used for the context manager:

Program code is given below:

```
class ram(object):
        def __init__(ayodhya, text):
        ayodhya.text = text
def __enter__(ayodhya):
        ayodhya.text = "You know  " + Ayodhya.text
        return ayodhya
def __exit__(ayodhya,exc_type,exc_value,traceback):
        ayodhya.text = ayodhya.text + "!"

with ram("Its a kingdom") as myram:
print(myram.text)


print(myram.text)
```

The output looks as follows:

You know: Its a kingdom!
You know: Its a kingdom!

When the object is initialized, the text property of the object is "Its a kingdom". As you can see, the object invokes the () and () methods as it enters and leaves the context, causing the text property of the object to change.

Pickle Pack

We can store the text in a file. But the most common objects in Python are objects that disappear from memory when the program ends, or the computer shuts down. Thus, can we save objects to disk?

You can do this with Python's pickle package. Pickle means pickle in English. Sailors at the Age of Discovery used to make pickles out of vegetables and take them with them in cans. Pickle in Python has a similar meaning. With the pickle package, we can save an object and store it as a file on disk.

In fact, objects are stored in two steps. The first step is to grab the object's data out of memory and convert it into an ordered text, called Serialization. The second step is to save the text to a file. When we need to, we read the text from the file and then put it into memory, we can get the original object. Here's a concrete example, starting with the first step of serialization, which converts an object in memory into a text stream:

```
import pickle
class Animal(object):
have_trunk = True
howtheyreproduce= "zygote"
winter= animal()
pickle_string = pickle.dumps(winter)
```

Using the pickle package's dumps() method, you can convert an object to a string. We then store the string in a file using the byte text storage method.

Step 2:

```
with open("winter.pkl", "wb") as f:
f.write(pickle_string)
```

The above procedure is deliberately divided into two steps to illustrate the whole process better. Instead, we can take a dump() approach and do two steps at a time:

```
import pickle
```

```
class Animal(object):
have_trunk = True
howtheyreproduce= "zygote"
winter= animal()
with open("winter.pkl", "w") as f:
pickle.dump(winter, f)
```

Object winter will be stored in the file winter (in the PKL). With this file, we can read the object if necessary. The process of reading an object is the opposite of that of storing it. First, we read the text from the file. Then, using the pickle's load() method, we convert the text as a string to an object. We can also combine the above two steps using the pickled load() approach.

Sometimes, just reversing the recovery is not enough. An object depends on its class, so when Python creates an object, it needs to find the appropriate class. So, when we read an object from the text, the class must already have been defined in the program. For built-in classes that Python always has, such as lists, dictionaries, strings, and so on, you don't need to define them in your program. For a userdefined class, however, you must first define the class before you can load its objects from the file.

Here is an example of a read object:

```
import pickle
class animal(object):
have_trunk = True
howtheyreproduce= "zygote"
with open("winter.pkl", "rb") as f:
winter= pickle.load(f)
print(winter.have_trunk)
```

Chapter 10: Object-Oriented Programming Part 1

Having looked at Python's process-oriented programming paradigm, we'll use a completely different programming paradigm in this chapter—object-oriented. Python is not just a language that supports an object-oriented paradigm. Under the multi-paradigm facade, Python uses objects to build its large framework. As a result, we can get an early start in object-oriented programming to understand Python's deep magic.

To understand object orientation, you need to understand classes and objects. Remember the process-oriented functions and modules that improve the reusability of your program. Classes and objects also improve program reusability. In addition, the class and object Syntax also enhances the program's ability to simulate the real world. "Emulation" is the very heart of object-oriented programming.

The object-oriented paradigm can be traced back to the Simula language. Kristen Nygaard is one of the coauthors of the language. He was recruited by the Norwegian Ministry of Defence and then served at the Norwegian Institute of Defence Sciences. As a trained mathematician, Kristen Nygaard has been using computers to solve computational problems in defense, such as nuclear reactor construction, fleet replenishment, and logistics supply, etc. To solve these problems, Nygaard needed a computer to simulate the real world. For example, such as what would happen if there was a nuclear leak. Nygaard found that, following a procedural, instruction-based approach to programming, it's hard for him to program real-world individuals. Take a boat, for example. We know it has some data, such as height Degree, width, horsepower, draught, etc. It will also have some movement, such as moving, accelerating, refueling, moored, and so on. This ship is one Individuals. Some individuals can be grouped, such as battleships and aircraft carriers are warships. Some individuals have inclusive relationships, such as A ship has an anchor.

When people tell stories, they naturally describe individuals from the real world. But for a computer that only knows the 0 / 1 sequence, it will just mechanically execute instructions. Nygaard hopes that when he wants to do computer simulations, it will be as easy as telling a story. He knew from his military and civilian experience that such a programming language had great potential. Eventually, he met Ole-Johan Dahl, a computer scientist. VEGETA is helping Nygaard turned his idea into a novel language -- Simula. The name of the language is the very simulacrum that Nygaard craves.

We can think of object orientation as a bridge between story and instruction. The coder uses a story-based programming language.

The compiler then translates these programs into machine instructions. But in the early days of computers, these extra translations consumed too many computer resources. Therefore, the object-oriented programming paradigm is not popular. Pure object-oriented languages are often criticized for their inefficiencies.

With the improvement of computer performance, the problem of efficiency is no longer a bottleneck. People turned their attention to the productivity of programmers and began to explore the potential of object-oriented languages. The first great success in the object-oriented world was the C++ language. Bjarne Stroustrup created the C++ language by adding object-oriented syntax to the C language. C++ is a mixture of C language features, so it looks very complex. Later versions of the Java language moved toward a more purely object-oriented paradigm and quickly became commercially successful. C++ and Java were once the most popular programming languages. Microsoft's later C# and Apple's ongoing support for OBJECTIVE-C were also typical object-oriented languages.

Python is also an object-oriented language. It's older than Java. However, Python allows programmers to use it in a purely procedural way, so its object-oriented heart is sometimes overlooked. One of Python's philosophies is that "everything is an object." Both the process-oriented Paradigm we saw in Chapter 3, and the functional programming we will see in the future, are, in fact, the result of special object simulations. Therefore, learning object-orientation is a key part of learning Python. Only by understanding Python's objects can we see the full picture of the language.

Classes

When it comes to finding objects, the first thing we look at is a grammatical structure called a class. The concept of class here is similar to the concept of "class" in our daily life. In everyday life, we put similar things into a group and give the Group A name. Animals, for example, have trunks in common and reproduce by zygote. Any particular Animal is based on the archetype of an Animal.

Here's how we describe animals in Python:

```
class Animal(object):
trunk= True
howtheyreproduce= "zygote"
```

Here, we define a class with the keyword class. The class name is Animal. In parentheses, there is a keyword object, which means something, that is, an individual. In computer languages, we refer to individuals as objects. A class does not. There can be more than one. Animals can include neighbor elephant, the tiger running over the horizon, and a small yellow chicken kept at home.

Colons and indenting describe the code that belongs to this class. In the block of programs that fall under this category, we define two quantities—one for the trunk, and the other for reproduction—called the attributes of the class. The way we define animals is crude. Animals are just "hairy things that reproduce." Biologists would probably shake their heads if they saw this, but we're taking our first steps into a simulated world.

In addition to using data attributes to distinguish categories, we sometimes also distinguish categories based on what these things can do. For example, birds can move. In this way, the bird is distinguished from the type of house. These actions have certain consequences—such as changes in position due to movement. Some of these "behavior" properties are called methods. In Python, you typically illustrate a method by defining a function inside a class.

```
class Animal(object):
    trunk= True
```

```
howtheyreproduce= "zygote"
def roar(self, sound):
print(sound)
```

We added a new method attribute to the animal, which is roar(). The method roar() looks a lot like a function. Its first argument is self, which refers to the object itself within the method, which I'll explain in more detail later. It should be emphasized that the first argument to a method must be self, which refers to the object itself, whether or not the argument is used. The rest of the parameter sound is designed to meet our needs, and it represents the content of the bird song. The method roar() does print out the sound.

Objects

We define classes, but as the function definition, this is still just a matter of building weapons. To use this sharp tool, we need to go deep into the object. By calling the class, we can create an object under the class. For example, I have an animal named winter. It's an object, and it belongs to an animal. We use the previously defined animal to generate this object:

winter= Animal()

Use this sentence to create an object and explain that winter is an object that belongs to Animals. Now, we can use the code already written in animal. As an object, winter will have the properties and methods of an animal. A reference to a property is made by a reference in the form of object.attribute. For example:

print(winter.howtheyreproduce)

In the above way, we get the reproductive pattern of winter's species.
In addition, we can call methods to get winter to do what the animal allows. For example:

When we call the method, we pass only one parameter, the string "Roarrrr". This is where methods differ from functions. Although we have to add this self parameter when defining a method of a class, self is only used inside the class definition, so there is no need to pass data to self when calling the method. By calling the roar() method, my winter can scream.

So far, the data describing the object has been stored in the properties of the class. A generic attribute describes a generic feature of a class, such as the fact that animals have trunks. All objects that belong to this class share these properties. For example, winter is an object of animals, so winter has trunks. Of course, we can refer to a class attribute through an object.

For all individuals within a class, individual differences may exist for certain attributes. For example, my winter is pink, but not all animals are pink. Let's take the human class. Sex is a property of a person—not all humans are male or female. The value of this property varies from object to object. Tom is an object of the human race, and the sex is male. Estella is also a human object; the sex is female.

Therefore, in order to fully describe the individual, in addition to the generic class attributes, we need the object attributes used to describe the personality.

In a class, we can manipulate the properties of an object through self. Now we extend the Animal Class:

class Animal(object):

```
def roar(self, sound):
print(sound)
def whatcolor(self, color):
self.color = color
winter= Animal()
winter.whatcolor("pink")
print(winter.color)
```

In the method set(), we set the object's property color with the self parameter. As with class attributes, we can pass objects. Property to manipulate object properties. Since the object attribute depends on self, we must operate on the class attribute within a method. Therefore, object properties cannot be assigned an initial value directly below the class, like class properties.

Python does, however, provide a way to initialize object properties. Python defines a series of special methods. In a very specific way, it's called the Magic Method. A programmer can set special methods in a class definition. Python deals with special methods in a particular way. For the class() method, Python calls it automatically every time an object is created.

Therefore, we can initialize object properties inside the () method:

```
class Animal(object):
def __init__(self, sound):
self.sound = sound
print("my roar is", sound)
def roar(self):
print(self.sound)
winter= Animal("Gurrrrr")
winter.roar()
```

In the above class definition, we show how the class is initialized using the roar() method. Whenever an object is created The def_init_() method is called when, for example, the summer object is created. It sets the sound property of the object. You can call this object property through self in the roar() method. In addition to setting object properties, we can also set the Add additional instructions in self(). These instructions are executed when the object is created. When a class is called, it can be followed by a Parameter list. The data put in here will be passed to the parameters of (). With the () method, we can initialize object properties at object creation time.

In addition to manipulating object properties, the self parameter also has the ability to call other methods of the same class within a method, such as:

```
class Animal(object):
def roar(self, sound):
print(sound)
def roarcontinous(self, sound, n):
for i in range(n):
self.roar(sound)
winter= Animal()
winter.roarcontinous("gurrr", 10)
```

In the method roar(), we call another method in the class, roar(), through self.

Successors

Subclasses

The category itself can be further subdivided into subcategories. Animals, for example, can be further divided into Amphibians and Reptiles. In object-oriented programming, we express these concepts through Inheritance.

```
class Animal(object):
```

```
trunk= True
howtheyreproduce= "zygote"
def roar(self, sound):
print(sound)
class Amphibian(Animal):
waytheywalk= "bywalk"
edible = True
class Reptile(Animal):
```

In the class definition, the parenthesis is Animal. This shows that Amphibian belongs to a subclass of Animal, namely Amphibian inherited from animal. Naturally, Animals are the father of Amphibians. Amphibians will have all of Animals attributes. Although we only declare winter as an amphibian, it inherits the properties of the parent class, such as the data attribute trunk, and the method attribute roar(). The New Reptile class also inherited from Animal. When you create a Reptile object, the object automatically has Animal properties.

Obviously, we can use inheritance to reduce the repetition of Information and statements in the program. If we define Amphibians separately,

Amphibian and reptiles, rather than animals, have to be entered into the amphibian and reptile definitions separately. The whole process can become tedious, so inheritance improves the program's reusability. In the most basic case, the object is inside the parentheses of the class definition. The class object is actually a built-in class in Python. It serves as the ancestor of all classes.

Classification is often the first step in understanding the world. We learn about the world by classifying all sorts of things. Ever since our human ancestors, we've been sorting. The 18th century was a time of great maritime discoveries when European navigators went around the world, bringing back specimens of plants and animals that had never been seen before. People are excited about the proliferation of new species, but they also struggle with how to classify them. Carl Linnaeus has proposed a classification system that paves the way for further scientific discoveries through the subordination of parent and child classes. Object-oriented language and its inheritance mechanism are just the conscious classification process of simulating human beings.

Attribute Overlay

As mentioned above, in the process of inheritance, we can enhance the functionality of a subclass by adding attributes in which the parent class does not exist. In addition, we can replace properties that already exist in the parent class, such as:

Amphibian is a subspecies of Animal. In Amphibian, we define the method roar(). This method is also defined in Animal. As you can see, the Amphibian calls its own defined roar() method instead of the parent class. In effect, it's as if the method roar() in the parent class is overridden by the namesake property in the child class.

By covering methods, we can radically change the behavior of subclasses. But sometimes the behavior of a subclass is an extension of the behavior of a parent class. At this point, we can use the super keyword to call methods that are overridden in the parent class, such as:

In the chicken roar() method, we use super. It is a built-in class that produces an object that refers to the parent class. Using super, we call the methods of the parent class in the method with the same name as the child class. In this way, the methods of a subclass can both perform related operations in the parent class and define additional operations of their own.

What You Missed Out on All Those Years
List Objects

We went from the original "Hey its Jesus Christ!" all the way to the object. As the saying goes, "Life is like a journey; what's important is the scenery along the way." In fact, the previous chapters have seen objects many times. However, at that time, the concept of the object hasn't been introduced yet. It's time to look back at all the people we missed.

Let's start with an acquaintance, a list in a data container. It's a class, and you can find the class name using the built-in function:

From the results returned, we know that a is a list type. In fact, a type is the name of the class to which the object belongs. Each list belongs to this class. This class comes with Python and is pre-defined, so it's called a built-in class. When we create a new table, we are actually creating an object of the list class. There are two other built-in functions we can use to investigate the class information further: Dir() and help(). The function Dir() is used to query for all attributes of a class or object. You can try this.

We have used the help() function to query the documentation for the function. It can also be used to display the class description document.

Returns are not only a description of the list class but also a brief description of its properties. By the way, making a class description document is similar to making a function description document. We just need to add the desired description in a multiline string under the class definition:

Pass in your program is a special Python keyword that says "do nothing" in this syntax structure. This keyword preserves the structural integrity of the program.

From the above query, we see that classes also have many "hidden skills." Some list methods, for example, return information about a list: The list is greatly enhanced by the invocation of a method. Seeing the list again from an object's point of view feels like a great party.

Tuples and String Objects

Tuples, like lists, are considered as sequences. However, tuples cannot change the contents. Therefore, Tuples can only be queried, not modified:

Strings are special tuples, so tuple methods can be executed.

Although strings are a type of Tuple, there are a number of ways that strings (strings) can change strings. This sounds like a violation of the immutability of Tuples. In fact, these methods do not modify the string object, but delete the original string, and create a new string, so there is no violation of tuple immutability. The following summarizes the methods for string objects.

STR is a string and sub is a substring of str. S is a sequence whose elements are strings. Width is an integer indicating the width of the newly generated string. These methods are often used for string processing.

Chapter 11: Object-Oriented Programming Part 2

We'll explore the meaning behind Python's "everything is an object." Many grammars—such as operators, element references, and built-in functions—actually come from special objects. Such a design satisfies both Python's need for multiple paradigms and its need for rich syntaxes, such as operator overloading and real-time features, with a simple architecture. In the second half of this chapter, we delve into important mechanisms related to objects, such as dynamic typing and garbage collection.

Operators

We know that list is the class of the list. If you look at the attributes of a list with Dir(list), you see that one of the attributes is addition(). Stylistically, addition() is a special method. What's so special about it? This method defines the meaning of the + operator for a list object. When two list objects are added, the list is merged. The result is a consolidated list:

>>>print([10,32,43] + [45,26,49])

This becomes [10,32,43,45,26,49]

Operators, such as +, -, and, and or are implemented in special ways, such as:

"def" + "uvw"
#This will become defuvw

The following was actually done:

"def".addition("uvw")

Whether two objects can be added depends first on whether the corresponding object has addition() method. Once the corresponding object has an addition() method, we can perform an addition even if the object is mathematically non-additive. Operators with the same function are simpler and easier to write than special methods. Some of the following operations can be tricky to write in a special way.

Try the following, see what it looks like, and think about the operator:

>>>(5).multiplication(4) # 5*4

```
>>>True.Or(False) # True or False
```

The special methods associated with these operations can also change the way they are performed. For example, lists cannot be subtracted from each other in Python. You can test the following:

```
>>>[3,5,7] - [5,7]
```

There is an error message that the list object cannot be subtracted, that is, the list does not define the "-" operator. We can create a subclass of the list and add a subtraction definition by adding() methods, such as:

```
class Addsubstraction(list):
def  substraction(self, b): a = self[:]
b = b[:]
while len(b) > 0:
element_b = b.pop()
if element_b in a: a.remove(element_b)
return a print(Addsubstraction([1,2,3]) - Addsubstraction([3,4]))
```

In the example above, the built-in function Len() returns the total number of elements contained in the list. The built-in function subtraction() defines the operation of "-" to remove elements from the first table that appear in the second table. So, the two Addsubstraction objects that we created can be subtracted. Even if the sub() method has already been defined in the parent class, the method in the child class overrides the method with the same name as the parent class when redefined in the child class. That is, the operator will be redefined.

The process of defining operators is useful for complex objects. For example, humans have multiple attributes, such as name, age, and height. We can define human comparison by age alone. So that you can, for your own purposes, take what wasn't there the operation in is added to the object. If you've been in military training, you've probably played a game of "turn left, turn right." When the drill master shouts commands, you must take the opposite action. For example, if you hear "turn to the left," you must turn to the right. In this game, the operators "turn left" and "turn right" are actually redefined.

Element References

Here are some common table element references:

li = [23, 42, 34, 49, 65, 86]

print(li[4])

When the above program runs to Li[4], Python finds and understands the [] symbol and calls the getitem() method.

li = [23, 42, 34, 49, 65, 86]
print(li.getitem(4))

Output will be:

65

Take a look at the following and think about what it corresponds to:

li = [23, 42, 34, 49, 65, 86]
li.setitem(4, 60) print(li)

Output will be:
23,42,34,49,60,86

Just a Small Example for Dictionary Datatype

thisisdictionary= {"first":1, "second":2}
thisisdictionary.__delitem__("first")
print(thisisdictionary)
#prints first=1

Implementation of Built-In Functions

Like operators, many built-in functions are special methods that call objects. For example:

len([34,24,35,89])
#This actually says how many numbers of elements are present

What it actually does is explained below in detail:

[34,24,35,89].__len__()

The built-in function len() also makes writing easier than _len_().

Try the following and think of its corresponding built-in function. These are called mathematical functions if said in the right way.

(-69).__abs__()   # prints the absolute value of the element
(2.3).__int__()   # prints the nearest integer value of the element

There are many built-in mathematical functions that help to make programming easier for inductive and statistical purposes.

Attribute Management

In inheritance, we mentioned the mechanism of property overwriting in Python. In order to understand property coverage, it is necessary to understand Python properties. When we call an attribute of an object, the attribute may have many sources. In addition to the attribute from the object and the attribute from the class, this attribute may be inherited from the ancestor class. Properties owned by a class or object are recorded in. This is a dictionary, the key is the attribute name, and the corresponding value is an attribute. When Python looks for properties of an object, it looks for them in the order of inheritance.

Let's look at the following classes and objects, the Chicken class inherits from the Bird class, and summer is an object of the Chicken class:

Below is the complex code:

```
class Animal(object):
    trunk= True
    def roar(self):
        print("some roaring")
class Amphibians(Animal):
    walk= False
    def __init__(self, age):
        self.age = age
    def roar(self):
        print("gurrr")
winter= Amphibian(4)
print("===> winter")
print(winter.__dict__)
print("===> Amphibian")
print(Amphibian.__dict__)
print("===> Animal")
print(Animal.__dict__)
print("===> result")
print(object.__dict__)
```

Here's our output (Please don't get confused because this is quite complex):

===> winter{'age': 4}

===> Amphibian {'walk': False, 'roar':, '__module__': '__main__', '__doc__': None, '__init__': }
===>Animal {'__module__': '__main__', 'roar':, '__dict__':, 'trunk': True, '__weakref__':, '__doc__': None}
===>result {'__setattr__':, '__reduce_ex__'

The order is based on proximity to summer's objects. The first part is about the properties of the summer object itself, the age. The second part is about the properties of the chicken class, such as the fly and _init_() methods. The third part is the Bird class attribute, such as feather. The last part belongs to the object class and has properties like this.

If we look at the properties of the object summer with the built-in Function Dir, we see that the summer object contains all four parts. In other words, the properties of objects are managed hierarchically. For all of the properties that object summer has access to, there are four levels: summit, Amphibian, Animal, and object. When we need to call a property, Python will iterate down layer by layer until we find the property. Because objects do not need to store the properties of their ancestor classes repeatedly, a hierarchical management mechanism can save storage space.

An attribute may be defined repeatedly at different levels. As Python traverses down, it picks the first one it encounters. This is how property coverage works. In the output above, we see that both Chicken and Bird have the roar() method. If you call the roar() method from winter, you will be using a version of Amphibian that is closer to the object winter.

winter.roar()

Properties of a subclass have precedence over properties of the same name of the parent class, which is the key to property overrides.

It is important to note that the above are all operations that invoke properties. If you do the assignment, then Python doesn't have to drill down into layers. Here's how to create a new Amphibian class object rainy, and how to modify attributes such as trunk with rainy:

rainy= Amphibian(3)
rainy.trunk = False
print(winter.trunk)

Although rainy modifies the trunk attribute value, it does not affect Animal's class attribute. When we look at rainy's object properties using the following method, we see that a new object property named trunk is created.

Print(rainy.__dict__)

Instead of relying on inheritance, we can directly manipulate the properties of an ancestor class, such as:

Animal.trunk = 3

Its equivalent to modifying Bird's:

Animal.__dict__["trunk"] = 3

Features

There may be dependencies between different properties of the same object. When a property is modified, we want other properties that depend on that property to change at the same time. At this point, we cannot store attributes in a static dictionary manner.

Python provides a variety of ways to generate attributes on the fly. One of these is called a property. A property is a special property. For example, we added an adult feature to the Amphibian class. When the age of the object exceeds 1, adult is true; otherwise, false.

Here is the programming code:

```
class Animal(object):
        trunk= True

 class Amphibian(Animal):
        walk = False
def __init__(self, age):
self.age = age

def adultage(self):
    if self.age > 1.0:
      return True
    else:
      return False

adultery = property(adultage) # property is built-in

winter= Amphibian(2)
print(winter.adultery) #True
```

```
winter.age = 0.5
print(winter.adult)   #False
```

The property is created using the built-in function property(). Property() can load up to four parameters. The first three arguments are functions that set what Python should do when it gets, modifies, and deletes features. The last parameter is a property of the document, which can be a string, for illustration purposes.

The upper num is a number, and the Neg is a property that represents the negative number. When a number is definite, its negative number is always definite. When we modify a negative number, the value of the number itself should also change. These two points are implemented by get() and set(). Del() indicates that if you delete the feature Neg, then the action that should be performed is Delete attribute value. The last parameter of property()("I'm negative") is the documentation for the feature neg.

getatr() method

In addition to the built-in function property, we can also use (self, name) to query for properties that are generated on the fly. When we call a property of an object, if the property cannot be found by mechanism, Python will call the() method of the object to generate the property immediately, such as:

Each feature needs its own handler, and() can handle all the instant generated properties in the same function. () can handle different properties depending on the function name. For example, when we queried for the attribute name male above, we threw an error of the AttributeError class. Note that () can only be used to query for properties that are not in the system.

(self, name, value) and (self, name) can be used to modify and delete attributes. They have a wider range of applications and can be used for any attribute.

Real-time attribute generation is a very interesting concept. In Python Development, you might use this approach to manage the properties of objects more reasonably. There are other ways to generate properties on the fly, such as using the descriptor class. Interested readers may refer to this code below.

```
class Animal(object):
trunk= True
class Amphibian(Animal):
walk = False
def __init__(self, age):
self.age = age
def __getattr__(self, name):
 if recognition== "old":
if self.age > 1.0:
 return True
else:
return False
else:
raise AttributeError(name)
winter = Amphibian(2)
print(winter.old) #True
winter.age = 0.5
 print(winter.adult) #False
 print(winter.male) # AttributeError
```

Dynamic Type

Dynamic Typing is another important core concept of Python. As I said earlier, Python variables do not need to be declared. When assigning a value, a variable can be reassigned to any other value. Python variables change from wind to wind. The ability of sand is the embodiment of dynamic type. Let's start with the simplest assignment statement:

c= 3

In Python, the integer 3 is an object. The object's name is "c". But more precisely, an object name is actually a reference to an object. An object is an entity stored in memory. But we don't have direct access to the subject. An object name is a reference to that object. Manipulating Objects by reference is like picking up beef from a hot pot with chopsticks. The object is beef, and the object name is the good pair of chopsticks.

With the built-in function ID(), we can see which object the reference points to. This function returns the object number.

c= 3
print(id(3))
print(id(c))

As you can see, after the assignment, object 3 and reference c return the same number.

In Python, an assignment is simply to use the object name as a chopstick to pick up other food. Each time we assign a value, we let the reference on the left point to the object on the right. A reference can point to a new object at any time:

c= 5
print(id(c))
c= "for"
print(id(c))

In the first statement, 3 is an integer object stored in memory. By assignment, the reference a points to object 5. In the second statement, an object "at" is created in memory, which is a string. The reference a points to "for". By returning the ID() twice, we can see that the object to which the reference is pointing has changed. Since the variable name is a reference that can be changed at any time, its type can naturally be changed dynamically in the program. Therefore, Python is a dynamically typed language.

A class can have more than one equal object. For example, two long strings can be different objects, but their values can be equal.

In addition to printing the ID directly, we can also use the is operation to determine whether two references point to the same object. But for small integers and short strings, Python caches these objects instead of frequently creating and destroying them. Therefore, the following two references point to the same integer object 5.

Mutable and Immutable Objects

With the first two statements, we let c and d point to the same integer object 5. Where c & d is meant to point the reference d to the object referred to by the reference c. We then manipulate the object, adding 3 to c, and assigning it to d. As you can see, a points to integer object 8, while d still points to object 5. Essentially, the addition operation does not change the object 5. Instead, Python just points a to the result of the addition -- Another object, 8. It's like a magic trick to turn an old man into a young girl. In fact, neither the old man nor the young girl has changed. It's just a girl on an old man's stage. In this case, it's just a reference point. Changing a reference is not Can Affect the direction of other references. In effect, each reference is independent and does not affect the other.

Below is the code:

```
c= 5
 print(id(c))
d= c
print(id(c))
print(id(d))
c= c+ 3
print(id(c))
print(id(8))
print(id(d))
```

When we changed LIST1, the contents of list2 changed. There seems to be a loss of independence between references. It's actually not a contradiction. Because the directions of LIST1 and LIST2 have not changed, they are still the same list. But a list is a collection of multiple references. Each element is a reference, such as list1[0], list1[1], and so on. Each reference points to another object, such as 23,45,76. And LIST1[0]10, the assignment, is not changing the direction of list1, but LIST1[0]. Hence, the direction of an element, which is part of the list object, changes. Therefore, all references to this list object are affected.

Therefore, when you manipulate lists, if you change an element by reference to an element, the list object itself changes (in-place change). Lists are objects that can change on their own, called Mutable objects. The dictionary we've seen before is also a variable data object. But previous integers, floating-point numbers, and strings cannot change the object itself. An assignment can only change the direction of the reference at most. Such objects are called Immutable objects. Tuples contain multiple elements, but these elements cannot be assigned at all, so they are immutable data objects.

Below is the code:

```
list2 = [23,45,76]
list1 = list2
```

```
list1[0] = 10
print(list2)
```

Look at the Function Parameter Passing from the Dynamic Type

The parameter x is a new reference. When we call function F, a is passed as data to the function, so x will point to the object referred to by a, which is an assignment. If a is an immutable object, then references a and x are independent of each other, meaning that the operation on parameter x does not affect references a.

In the function above, a points to a variable list. When the function is called, a pass the pointer to the parameter X. At this point, both references to a and x point to the same mutable list. As we saw earlier, manipulating a mutable object by a reference affects other references. The results of the program run also illustrate this point. When you print a, the result becomes [100,2,3]. That is, the action on the list inside the function is "seen" by the external reference A. Be aware of this problem when programming.

Below is the code:
```
def f(x):
```

```
print(id(x))
x = 50
print(id(x))
a = 2
print(id(a))
f(a)
print(a)
```

Memory Management in Python

1. Reference Management

Language memory management is an important aspect of language design and an important determinant of language performance. Whether it's C Manual language management or Java garbage collection, is the most important feature of the language. Take the Python language as an example to illustrate how memory is managed in a dynamically typed, object-oriented language.

First, let's make it clear that object memory management is based on the management of references. We've already mentioned that in Python, references are separated from objects. An object can have multiple references, and each object has a total number of references to that object, the Reference

Count. We can use getrefcount() in the sys package in the standard library to see the reference count of an object. Note that when you pass a reference to getrefcount() as a parameter, the parameter actually creates a temporary reference. Therefore, getrefcount() gets more than 1.

2. Garbage Collection

If you eat too much, you'll get fat, and so will Python. As the number of objects in Python increases, they take up more and more memory. But you don't have to worry too much about Python's size, as it will be smart enough to "lose weight" and start recycling in due course Garbage Collection, which purges objects of no use. Garbage collection mechanisms exist in many languages, such as Java and Ruby. While the ultimate goal is to be slim, weight loss programs vary greatly from language to language.

In principle, when the reference count of an object in Python drops to zero, meaning that there are no references to the object, the object becomes garbage to be recycled. For example, if a new object is assigned to a reference, the reference count of the object becomes 1. If the reference is deleted and the object has a reference count of 0, then the object can be garbage collected.

The code is below:

variable= [12, 24, 36]
del variable

After del variable, there is no reference to the previously created table [12,24,36], which means that the user can not touch or use the object in any way. If this object stays in memory, it becomes unhealthy fat. When garbage collection is started, Python scans the object with a reference count of 0, emptying the memory it occupies.

Weight loss, however, is an expensive and laborious process. When garbage is collected, Python cannot perform other tasks. Frequent garbage collection can drastically reduce Python's productivity. If there are not many objects in memory, it is not necessary to start garbage collection frequently. Therefore, Python will only automatically start garbage collection under certain conditions. When Python runs, the number of times an Object Allocation and an Object Deallocation are logged. Garbage collection starts when the difference between the two is above a certain threshold.

In addition to the basic recycling approach described above, Python also employs a Generation recycling strategy. The basic assumption of this strategy is that objects that live longer are less likely to become garbage in later programs. Our programs tend to produce large numbers of objects, many of which are quickly created and lost, but some of which are used over time. For reasons of trust and efficiency, we believe that such "long-lived" objects can still be useful, so we reduce the frequency with which they are scanned in garbage collection.

Chapter 12: Exception Handling

This chapter deals with debugging and exception handling in detail. First of all, we will start with a small introduction about a bug to get a good overview of the topic.

What Is a Bug?

Bugs must be the most hated creatures a programmer can have. A bug in the programmer's eyes is a bug in a program. These bugs can cause errors or unintended consequences. Many times, a bug can be fixed after the fact. There are, of course, irremediable lessons. The European Ariane 5 rocket exploded within a minute of its first launch. An after-action investigation revealed that a floating-point number in the navigator was to be converted to an integer, but the value was too large to overflow. In addition, a British helicopter crashed in 1994, killing 29 people. The investigation revealed that the helicopter's software system was "full of flaws." In the movie 2001: A Space Odyssey, the supercomputer HAL kills almost all of the astronauts because of two goals in its program conflict.

In English, bug means defect. Engineers have long used the term bug to refer to mechanical defects. And there's a little story about using the word bug in software development. A moth once flew into an early computer and caused a computer error. Since then, bugs have been used to refer to bugs. The moth was later posted in a journal and is still on display at the National Museum of American History.

Code:

```
for result in range(5)
print(result)
# Python does not run this program. It will alert you to grammatical errors:
```

Output is:

SyntaxError: invalid syntax

There are no syntax errors in the following program, but when Python is run, you will find that the subscript of the reference is outside the scope of the list element.

```
result= [12, 24, 36]
print(result[4])

# The program aborts the error reporting
```

Output:

IndexError: list index out of range

The above type of Error that the compiler finds only at Runtime is called the Runtime Error. Because Python is a dynamic language, many operations must be performed at run time, such as determining the type of a variable. As a result, Python is more prone to run-time errors than a static language.

There is also a type of Error called a Semantic Error. The compiler thinks that your program is fine and can run normally. But when you examine the program, it turns out that it's not what you want to do. In general, such errors are the most insidious and the most difficult to correct. For example, here's a program that prints the first element of a list.

```
mix = ["first", "second", "third"]
print(mix[1])
```

There is no error in the program, normal print. But what you find is that you print out the second element, B, instead of the first element. This is because the Python list starts with a subscript from 0, so to refer to the first element, the subscript should be 0, not 1.

Debugging

The process of fixing a bug in a program is called debugging. Computer programs are deterministic, so there is always a source of error. Of course, sometimes spending a lot of time not being able to debug a program does create a strong sense of frustration, or even a feeling that you are not suitable for program development. Others slam the keyboard and think the computer is playing with itself. From my personal observation, even the best programmers will have bugs when they write programs. It's just that good programmers are more at peace with debugging and don't doubt themselves about bugs. They may even use the debug process as a kind of training, to work with their computer knowledge by better understanding the root cause of the error.

Actually, debugging is a bit like being a detective. Collect the evidence, eliminate the suspects, and leave the real killer behind. There are many ways to collect evidence, and many tools are available. For starters, you don't need to spend much time with these tools. By inserting a simple print() function inside the program, you can see the state of the variable and how far it has run. Sometimes, you can test your hypothesis by replacing one instruction with another and seeing how the program results change. When all other possibilities are ruled out, what remains is the true cause of the error.

On the other hand, debug is also a natural part of writing programs. One way to develop a program is Test-Driven Development (TDD). For Python to be such a convenient, dynamic language, it's a good place to start by writing a small program that performs a specific function. Then, on the basis of the small program, gradually modify, so that the program continues to evolve, and finally, meet the complex requirements. Throughout the process, you keep adding features, and you keep fixing mistakes. The important thing is, you've been coding. The Python author himself loves this kind of programming. So, debug is actually a necessary step for you to write the perfect program.

Exception Handling in Detail

For errors that may occur at run time, we can deal with them in the program in advance. This has two possible purposes: one is to allow the program to perform more operations before aborting, such as providing more information about the error. The other is to keep the program running after it makes a mistake.

Exception handling can also improve program fault tolerance. The following procedure uses the exception handling:

The program that requires exception handling is wrapped in a try structure. Except explains how the program should respond when a particular error occurs. Program, input() is a built-in function to receive command-line input. The float() function is used to convert other types of data to floating-point numbers.

If you enter a string, such as "P", it will not be converted to a floating point number, and trigger ValueError, and the corresponding except will run the program that belongs to it. If you enter 0, then dividing by 0 will trigger ZeroDivisionError. Both errors are handled by the default program, so the program does not abort.

The complete syntax for exception handling is:

try:
... (code should be written here)
except exception1:
... (code should be written here)
except exception2:
... (code should be written here)
else:
... (code should be written here)
finally:
...

If an exception occurs within a try, the exception is assigned, and except is executed. Exception layer by layer to see if it is exception1, exception2, and so on, until it is found to belong to, execute the corresponding statements in except. If there is no exception in try, So the except part skips the execution of the else statement. Finally is something you do in the end, whether or not there is an exception. If except is followed by no parameters, then all exception will be handled by the program.

Chapter 13: Python Web Programming

This chapter briefly explains web programming with Python. The Internet is a basic entity for crores of people now. Explanation of web modularity with Python can help you to learn the subject better. However, we will just go through the basics for now. For advanced web programming, please follow books with precise information.

HTTP Communication Protocol

Communication is a wonderful thing. It allows information to be passed between individuals. The animals send out the chemical element and mating messages. People say sweet things to express their love to their lovers. The hunters whistled and quietly rounded up their prey. The waiter barked to the kitchen for two sets of fried chicken and beer. Traffic lights direct traffic, television commercials broadcast, and the Pharaoh's pyramids bear the curse of forbidden entry. With communication, everyone is connected to the world around them. In the mysterious process of communication, the individuals involved always abide by a specific protocol. In our daily conversation, we use a set grammar. If two people use different grammars, then they communicate with different protocols, and eventually, they don't know what they're talking about.

Communication between computers is the transfer of information between different computers. Therefore, computer communication should also follow the Communication Protocol Conference. In order to achieve multi-level global Internet communication, computer communication also has a multi-level protocol system. HTTP Protocol is the most common type of network protocol. Its full name is the Hypertext Transfer Protocol or Hypertext Transfer Protocol.

The HTTP protocol enables the transfer of files, especially hypertext. In the Internet age, it is the most widely used Internet Protocol. In fact, when we visit a Web site, we usually type an HTTP URL into the browser, such as http://www.google.com, for example, says that you need to use the HTTP protocol to access your site.

HTTP works as a fast food order:

1. REQUEST: A customer makes a request to the waiter for a chicken burger.
2. Response: The server responds to the request of the customer according to the situation.

Depending on the situation, the waiter may respond in a number of ways, such as:

- The waiter prepares the DRUMSTICK Burger and hands it to the customer. (everything is OK)
- The waitress found herself working at the dessert stand. He sent his customers to the official counter to take orders. (redirects)
- The waiter told the customer that the drumstick hamburger was out. (cannot be found)

When the transaction is over, the waiter puts the transaction behind him and prepares to serve the next customer.

GET /start.html HTTP/3.0
Host: www.mywebsite.com

In the starting line, there are three messages:

- Get method. Describes the operation that you want the server to perform.
- / start. The path to the html resource. This points to the index on the server. HTML file.
- HTTP 3.0. The first widely used version of HTTP was 3.0, and the current version is 3.3.

The early HTTP protocol had only the GET method. Following the HTTP protocol, the server receives the GET request and passes the specific resource to the client. This is similar to the process of ordering and getting a Burger from a customer. In addition to the GET method, the most common method is the POST method. It is used to submit data from the client to the server, with the data to be submitted appended to the request. The server does some processing of the data submitted by the POST method. The sample request has a header message. The type of header information is Host, which indicates the address of the server you want to access.

After receiving the request, the server will generate a response to the request, such as:

```
HTTP/3.0 200 OK
Content-type: text/plain
Content-length: 10
Jesus Christ
```

The first line of the reply contains three messages:

- HTTP 3.0: Protocol version
- 200: Status Code
- Ok: Status Description

Ok is a textual description of the status code 200, which is just for human readability. The computer only cares about three-digit status codes. Status Code, which is 200 here. 200 means everything is OK, and the resource returns normally. The status code represents the class that the server responded to.

There are many other common status codes, such as:

- 302, Redirect: I don't have the resources you're looking for here, but I know another place where xxx does. You can find it there.

- 404, Not Found: I can't find the resources you're looking for.

The next line, Content-type, indicates the type of resource that the body contains. Depending on the type, the client can start different handlers (such as displaying image files, playing sound files, and so on). Content-length indicates the length of the body part, in bytes. The rest is the body of the reply, which contains the main text data.

Through an HTTP transaction, the client gets the requested resource from the server, which is the text here. The above is a brief overview of how the HTTP protocol works, omitting many details. From there, we can see how Python communicates with HTTP.

http.client Package

The client package can be used to make HTTP requests. As we saw in the previous section, some of the most important information for HTTP requests are the host address, request method, and resource path. Just clarify this information, plus http. With the help of the client package, you can make an HTTP request.

Here is the code below in Python:

import http.client
connection = http.client.HTTPConnection("www.facebook.com") #hostaddress conn.request("POST", "/") # requestmethod and resource path
response = connection.getresponse() # Gets a response
print(response.status, response.reason)# Replies with status code and description
content = response.read()

Conclusion

Thank you for making it through to the end of *Learn Python Programming*! Let's hope it was informative and able to provide you with all of the tools you need to achieve your goals—whatever they may be.

The next step is to get practice with Python in detail. Remember that programming is not an easy job. You need to master basics and use them to build a house of knowledge that is well-structured and organized. You may often get irritated with errors. However, always motivate yourself to work hard. Try to utilize internet resources like stack overflow for increasing your productivity if you are struck with any code.

Python is a great programming language for beginners due to its extensive resources and projects in GitHub for beginners. Try to contribute back to the Python community with all of your strength. Now, go program!

Finally, if you found this book useful in any way, a review on Amazon is always appreciated!